Problematising Water

Problematising Water: An Experiment on the Edges of Interdisciplinarity is a manifestation of thinking with water across a range of disciplines including human geography, sociology, anthropology, science and technology studies, and ethnomethodology.

In this book, Vitellone reveals the alliances between Michel Foucault's concept of problematisation and a new methodological approach for water-focused research, as seen through her study of London's #OneLess refill water fountain pilot, an intervention to reduce plastic bottled water. Situating the refill water fountain as a problematic space for collaborating with marine scientists, marine conservationists, systems change practitioners, and publics engaged in the making and doing of refill, the book discusses how Foucault's notion of problematisation forces a re-thinking of knowledge making practices in environmental research on water. It also synthesises a large body of literature by key thinkers, including Bruno Latour, Isabelle Stengers, Andrea Ballestero, Astrida Neimanis, Noortge Marres, Gay Hawkins, Andrew Barry, Mike Michael, and others, and links to current debates across a variety of disciplines.

Problematising Water: An Experiment on the Edges of Interdisciplinarity is a timely and provocative contribution to debates around the climate crisis and environmental interventions and the role that different research methods and objects play within them. It will be of value to science and technology scholars, social scientists, natural scientists, and academics interested in the environmental humanities.

Nicole Vitellone is AF Warr Senior Lecturer in Sociology in the School of Law and Social Justice and Co-Director of Centre for Health, Arts, Society & Environment (CHASE) at the University of Liverpool, United Kingdom.

Routledge Explorations in Environmental Studies

China and Gulf Cooperation Council Countries
Partnerships Towards Economic and Energy Transition
Edited by Dongmei Chen, Fuqiang Yang, Lei Yang and Majed Alsuwailem

Voluntary Environmental Programs in the United States and China
Pianpian Wang

Planning Law and Environmental Limits
A New Legal Architecture for Emissions
Naomi Luhde-Thompson

A Critical Introduction to the Economics of Sustainability
Brett Caraway

Problematising Water
An Experiment on the Edges of Interdisciplinarity
Nicole Vitellone

Designing Multispecies Neighbourhoods in Urban Seascapes
Placemaking and Storytelling in Scandinavian Environments
Edited by Elin Tanding Sørensen

For more information about this series, please visit: www.routledge.com/Routledge-Explorations-in-Environmental-Studies/book-series/REES

Problematising Water

An Experiment on the Edges of Interdisciplinarity

Nicole Vitellone

Routledge
Taylor & Francis Group
LONDON AND NEW YORK

First published 2026
by Routledge
4 Park Square, Milton Park, Abingdon, Oxon OX14 4RN

and by Routledge
605 Third Avenue, New York, NY 10158

Routledge is an imprint of the Taylor & Francis Group, an informa business

British Library Cataloguing-in-Publication Data
A catalogue record for this book is available from the British Library

ISBN: 9781041018766 (hbk)
ISBN: 9781041018780 (pbk)
ISBN: 9781003616757 (ebk)

DOI: 10.4324/9781003616757

Typeset in Times New Roman
by Newgen Publishing UK

Contents

Acknowledgements *vi*

Introduction: How to think, research, and write with water – a problematological turn 1

1 Thinking problematically with the water fountain 29

2 Probing problematisation in the plastics and climate crises 47

3 Loitering as a method of interdisciplinary research 65

4 Sociography of a curious object 88

5 Using ethnomethodology to make sense of the #OneLess refill water fountain pilot 109

Index *137*

Acknowledgements

This monograph was supported by research leave and funding from the School of Law and Social Justice (SLSJ) at the University of Liverpool. I'm grateful to SLSJ for supporting blue-skies research and members in LSRC (Liverpool Sociological Research Cluster) and CHASE (Centre for Health, Arts, Society and Environment) for being great colleagues. Special thanks to Michael Mair, Phil Brooker, Paul Jones, Gabe Mythen, Andrew Kirton, Louise Hardwick, Di Frost, Kirsty Morrin, Pete Campbell, Ross McGarry, Lena Theodoropoulou, Adele Moore, Ciara Kierans, Sally Sheard, Sam Solnick, Bethan Evans, Josie Billington, Jacquie Waldock, Mette Kragh-Furbo, Georgia Petridou, Stephen Kenny, and Claire Pierson. My thinking for this project has especially benefited from the intellectual instruction and ongoing conversations with Ciara Kierans, Michael Mair, and Phil Brooker. I'm so grateful for their methodological contributions and ethnomethodological insights, especially in the final stages where they read and commented on the monograph and made me see things more clearly. The writing up and completion of this book has greatly benefited from a Visiting Senior Research Fellowship Award from NICHE Centre for Environmental Humanities at Ca' Foscari University of Venice. Special thanks to the Director, Francesca Tarocco, and members of NICHE for providing an intellectually stimulating space and provocations on how to think with water that widened and sharpened the interdisciplinary focus of this project. Chapters in this book have benefited from the feedback of peers at conferences, seminars and journals, and the contribution of students on my module 'Using Foucault's Methods'. I want to thank my postgraduate students; co-panelists at 2024 EASST/STS session on water and experimental articulations of knowledge, especially the discussant Christy Spackman; the Waterscapes Seminar Series at NICHE; and Jane Kilby and Graeme Gilloch for their editorial feedback on an earlier version of Chapter 4 which appeared in the special issue 'Sociography: Writing differently'. I am grateful to the publisher (Sage) for their permission to reproduce this article 'Sociography of a curious object: descriptive research, the #OneLess refill pilot and practices of hesitation', *The Sociological Review* 2022 70(4): 666–685. My greatest thanks go to the #OneLess Marine and Freshwater Team in the Conservation and Policy Department of the Zoological Society of London (ZSL) and the stakeholders and research participants whose time, input, and knowledge were invaluable in making this project possible. Special thanks to Heather Koldeway and Fiona Llewellyn at ZSL and Lydia Clarkson. Big thank you to colleagues and friends who sustained my focus over the years: Monica Pearl, Joanna Hodge, Roberta Mazza, Shirley Tate, Vicky Singleton, Jackie Stacey, Ana Carden-Coyne, Helen Keane, Steven Angelides, Liz Wilson, Vicki Kirby, Cameron Duff, Justin O'Conner, Dave Whyte; my mates Franci, Nunzia, Rob, Marco, Scottie, Debs, Stuart, Mr Minch, John, Rhiannon, Ash, Ian, Davide, Colin, Meredith, Antonella, Eleanor, Debbie, Helen, Julie, Jo for their kindness and joy; my family Antonietta D'Angelica, Vincenzo Vitellone, Marianne, Molly and April Dodd for their love and support, and last but not least, Jayne Compton whose unwavering support and creative flair knows no limits.

Introduction

How to think, research, and write with water – a problematological turn

In the last couple of decades, the overt environmental and ecological damage sparked by the plastics pollution crisis has gradually become normalised, and there is now an acknowledgement of the impact of the marketing of bottled plastic water on how we drink and think about water. In discussing what makes the present moment different in the history of water in the urbanised industrialised west, the historian Christopher Hamlin (2000) points out the tensions in the ways we think about water as a concept. It is a conflict he argues between two distinct ways of conceptualising water. The first, is the monolithic conceptualisation of water presented by public water utilities. The second approach involves the marketing of bottled water to consumers as many by the bottled water industry. In drawing attention to the ways in which the concept of water has become a *contested concept*, Hamlin (2000) argues public waterworks managers must reconceptualise waters as diverse, with a focus on taste and quality, in order for the public to switch from bottled water back to the tap. Recognising the importance of waters as a concept, argues Hamlin, 'may have significant impact on public assessment of water technologies' (p. 313). This is why, he argues, the conceptualisation of water as one or many has implications for contemporary water policy. Depending on which of these concepts becomes ascended will directly impact public water provision in urbanised industrialised contexts, with the possible scenario of 'a two-tier water society: those who buy bottles and those who take from the terrible tap' (p. 322). At the root of Hamlin's contemporary history of water as a contested concept, is a recognition that changing conceptualisations of water and water technologies have social and political significations, and that 'people think – and have thought – in more complicated ways about the central question of what water is and about what the criteria are with

DOI: 10.4324/9781003616757-1

which one ought to assess it' (p. 315). Historians who write about the concept, Hamlin suggests, 'may have a role to play in which of these scenarios prevails' (p. 323).

In response to this paradigm shift from water to waters, some scholars have begun to pursue more radical responses, launching initiatives like discard studies and bringing terms such as 'pollution' to the forefront of environmental discussion on plastics. In *Discard studies: Social studies of waste, pollution and externalities*, the plastic pollution scholar Max Liboiron with Josh Lepawsky (2022) propose an interdisciplinary methodology for disrupting the framing of plastics as an environmental problem of individual consumers, and understanding the role of plastics as pollution and colonialism (see Liboiron 2021). The turn to investigating plastic production in the social sciences, is evidenced by the title of Alice Mah's (2022) book *Plastic unlimited: How corporations are fuelling the ecological crisis and what we can do about it*. These works respond not just to the current environmental moment on plastic pollution, but to methodological and epistemological challenges on how academics respond to problems and challenge how we think. Importantly, these works do not simply ask academics and social researchers to switch the focus from plastic consumption to plastic production, but reassess how this distinction is constituted and its political effects for normalising the ecological plastics crisis as beyond dispute. Although Liboiron and Lepawsky (2022) and Mah (2022, 2023) have a clear preference for focusing on plastic production, their intervention of questioning the global plastics industry, including challenging the continual growth in single-use plastics by global corporations and the production of plastic waste and recycling, undermines the neat distinction between consumption and production in the conceptualisation of plastic bottled water. While some analyses of plastic occasionally address both production and consumption, as with Gay Hawkins' (2011, 2013) account of plastic water (discussed in Chapters 1 and 2), what often goes unnoticed in science and technology informed studies and social scientific research are the specific ways the public respond to such debates about water. This book attempts to fill this gap. By turning to my empirical study of London's #OneLess refill water fountain pilot, I offer a mode of engagement that teaches us something about how people think with water and its status as a contested concept.

The #OneLess campaign ran in London from January 2016 to January 2022. It was established by the Marine CoLABoration as an experiment to shift single-use plastic water bottles. Based in London, the campaign was led by marine scientists and marine conservationists at the Zoological Society of London (ZSL), in conjunction with the Mayor of London, to reduce plastic bottled water consumption. The campaign involved a public refill water fountain pilot, which aimed to distribute free drinking water and change the way Londoners drink water. During 2018 and 2019, I conducted fieldwork on sites across London where the pilot took place. My study involved an empirical investigation of the human and non-human actors and devices of environmental participation involved in the #OneLess refill experiment. My empirical engagement with the public refill water fountain and the publics' interactions with it involved the use of research methods of close observation, thick description, and interviewing to deepen understanding of how environmental knowledge of plastic pollution becomes a matter of concern in everyday practices. My study of the mundane technology of the water fountain involved a collaboration with the marine scientists and conservtionists at ZSL and #OneLess systems change practitioners and stakeholders responsible for installing and maintaining the refill water fountain in three sites. The objective of my research was to address what's at stake in London's #OneLess experiment and investigate the effects of the refill fountain to reduce single-use plastic bottled water. As we will see in Chapter 3, this question also concerns what is at stake in the process of conducting research on everyday practice, and the methodological issues that arise when academics from different disciplines collaborate for the shared purpose of evaluation. This opens up the more general question of how the water fountain created a space for collaboration and interdisciplinary encounters across the natural and social sciences.

In seeking to begin to understand what is at stake in my case study of the water fountain, Chapters 1 and 2 address the reaction of social scientists and humanities scholars to the climate and plastics crises. By contextualising studies of environmental problems of plastic bottled water in various theoretical debates and methodological developments, the chapters show how different academics from divergent disciplines have engaged with the public water fountain, and the capacities and limits of different methods and research techniques to understand how people think with water. Chapters 3, 4, and 5 present original findings from my empirical study of the #OneLess refill water fountain pilot.

Focusing on the use of three refill water fountains in their various settings, these chapters show what an investigation of the water fountain has to offer for understanding professional and everyday modes of participation in environmental experiments to produce problematic modes of thinking. Offering an empirical study of London's refill pilot as a mundane site of action and transformation, we probe new ways of addressing the process of thinking with water in practical settings and everyday practices through which understandings of water materialise. Rather than think about bottled drinking water, and study it's use as a problem or solution to urgent global environmental issues, the aim of *Problematising water: An experiment on the edges of interdisciplinarity* is to use the refill pilot as an entry point to address shared ways of responding to water troubles in everyday spaces and engage the water fountain as a mundane site of plastic waters contestation.

In the following sections of this introductory chapter, I situate my study of the water fountain more explicitly within the contemporary context, including conceptual accounts and empirical research on water, before offering an alternative engagement inspired by Michel Foucault's analysis of problematisation and ethnographic and ethnomethodological ways of thinking. Providing an overview of the ways water has been engaged with from different epistemological and methodological perspectives is an opening for the reader of the different ways thinking with water in humanities and social science disciplines have developed in response to global environmental problems such as the plastics crisis and access to free public drinking water. This overview will serve as a re-framing of the question on how to think, research, and write with water through a theoretical and methodological framework of problematisation that situates the water fountain as a key object for opening a space to interrogate lay and professional knowledge practices. In opening a space to engage with the water fountain as a case study to test and advance concepts and methods, the chapter foregrounds the role other academics have played in responding to societal challenges relating to environmental issues concerning water. In so doing, I make explicit the relevance of different research methods and techniques in water-focused studies as methodological devices for engaging in environmental troubles and issues in local sites. In drawing attention to the different disciplinary and interdisciplinary methods for researching water controversies, I introduce readers to the novel methodological and conceptual approaches others have developed to inform knowledge practices, and then propose an alternative approach.

The trajectory of theoretically engaged research on water in the social sciences and environmental humanities has produced important insights that prompt us to look at water as a contested concept and reality. As the scholar Jamie Linton observes in his 2010 book *What is water?* (discussed in Chapter 1), the conceptualisation of water has become part of scholarly and everyday life, both materially and mentally. As we will see in Chapter 1, Linton's epistemological and methodological contributions to water studies involve using the mundane water fountain as a case study to anchor his thinking. In this introductory chapter, we focus on scholars who have explicitly oriented their work to drawing attention to seen and unseen water in everyday life as sites of troubles and topics of study. Such observations have provoked thinking with water from a range of perspectives in the disciplines of geography (Bakker 2003; Swyngedouw 2004; Kaika 2005; Linton 2010; Linton and Budds 2014; Ekers and Loftus 2008), anthropology (Strang 2004, 2005; Kaplan 2011; Hastrup and Hastrup 2015; Anand 2017; Jensen 2017; Lora-Wainwright 2017; Ballestero 2019; Pia 2024), history (Hamlin 1992, 2000; Sheard 1994; Taylor and Trentmann 2011; Gentlicore 2023) and interdisciplinary perspectives informed by feminist cultural theory on environmental change (Chen, Macleod and Neimanis 2013; Neimanis 2014, 2019) and science and technology studies (STS) (de Laet and Mol 2000; Sofoulis 2005; Lavau 2013; Bijker 2007; Barnes and Alatout 2012; Hawkins, Potter and Race 2015; Spackman and Burlingame 2018; Spackman 2023; Whatmore and Landström 2011, Whatmore 2013a,b) involving a range of methodological practices including the sensory science of taste (Spackman 2023), an ethnography of devices (Ballestero 2019), and experimental research apparatuses (Whatmore and Landström 2011). My goal here is not to take a critical approach to the broad field of water-focused research but to capture what is most notable about these studies: their capacity to articulate nuanced methodologies about *how* to construct the meaning of water. I will briefly sketch out the work of Astrida Neimanis (2014, 2019), Andrea Ballestero (2019), Christy Spackman (2023), and Sarah Whatmore and Catharina Landström (2011) as examples that offer unusual descriptions of water and provide new visions for theory, research, and policy. I show how these authors provide new methodological visions for capturing the concept of water from a range of vantage points. I also suggest how they illuminate a major gap in scholarly attention to water and environmental problems: how thinking is done by people and professionals, how thinking with water intersects with knowledge on environmental issues including the plastics and climate crises, and how such thinking translates into contesting the concept of plastic bottled waters. The work of Michel Foucault on problematisation that I address throughout this book asks us not to dismiss this process. By investigating London's #OneLess refill water pilot as an important site for action and transformation, the book makes the water fountain a case study to situate a focus on problematisation and ground theory and methodological debates to mundane sites for reflection and insight. As we will see, attending to how thinking with water in everyday drinking practices is a matter of concern and contestation requires engaging with the water fountain in specific settings through methods of observation, interviewing, and ethnomethods.

How to think with water phenomenologically

Thinking *with* water is a methodological and conceptual framework that interrogates the limits of thinking about, or of water as an abstract concept. It combines feminist posthumanist theory with an interest in situating ourselves in relation to water, acknowledging our participation enabled by it in daily practices. Thinking with water, first emerged as a distinct methodological framework in the mid-2010s in two academic publications. The first was an edited collection *Thinking with water* (2013) which featured an introduction by the editors Cecilia Chen, Janine MacLeod, and Astrida Neimanis. The second was Neimanis' 2014 publication 'Alongside the right to water, a posthumanist feminist imaginary' in the *Journal of Human Rights and the Environment*. One of Neimanis' (2014) foundational insights is that water and our relations with it have been opposed in academic and public discourses on the human right to water. Neimanis troubles this opposition by taking up a set of posthumanist feminist concepts – naturecultures (Kirby 2008) and intra-action (Barad 2006) – as orientations interested in 'problematising the bifurcation of human and non-human nature' (p. 14). Whilst a posthumanist feminist perspective is seen to 'provide theoretical resources for engendering a more capacious aquatic imaginary' (Neimanis 2014, p. 13), which bring us closer to the matter of water and enable us to think alongside it, this new configuration of thinking with water as a relational concept is 'much more than a theoretical meander' (Chen, MacLeod and Neimanis 2013, p. 20). In *Bodies of water*, Neimanis (2019) refers to a methodology with a specific name – posthuman feminist phenomenology – as an approach for imagining water through lived embodied experience. The phenomenological practice of knowledge creation involves deep description of bodily experience, achieved through the situatedness of the practitioner in posthuman knowledge practices in the world. What I am specifically interested in here, is how Neimanis' example of drinking a glass of water is used as a posthuman phenomenological exercise to distinguish between two different types of waters – watery embodiment and the abstract concept of water:

> When I drink a glass of water, where does it go? How does is animate me? Where does it come from? How does water ask me to move, what sensations does it evoke, what effect does it engender, in me and the bodies it connects me to?
>
> (p. 24)

As a tool for thinking relationally with water, Neimanis' posthuman phenomenological method is understood to 'loosen what we know and open to what we do not' (p. 42). Reading this example reveals how description and thinking phenomenologically through the embodied sensorial practice of drinking are circumscribed by the *glass* of water. And it reminds us that we should remain attentive to the everyday material objects that situate thinking with water and for what they can enable in contesting plastic bottled water. More than just making explicit the embodied practice of drinking a glass of water as an empirical example of sense making, Neimanis' posthuman phenomenological method offers insights that prompt us to take seriously everyday engagements with the mundane technology of the water fountain as an embodied sensorial activity.

How to research water ethnographically

A future history of water by Ballestero (2019) features a significant material and technological component, probing the assumptions in scholarship on water of a neat distinction between water as a commodity and water as a human right. Using a descriptive methodological approach, Ballestero questions the tendency to reproduce this dominant distinction in conceptualisations of water. She demonstrates instead, how an ethnographic experiment less strictly focused on thinking with conventional water bodies allows us to imagine additional or different meanings about the materiality of water. Instead of focusing on humans or water, Ballestero (2019) investigates how meanings of water by everyday people in Costa Rica and Brazil are created relationally through ordinary bureaucratic and technical devices that shape everyday routines. Ballestero's ethnographic approach of thinking with device, 'as an alternate location' (p. 192) to begin her analytical work, challenges how we trace water itself 'beyond pipes, dams, rivers and oceans', and shifts the focus of inquiry away 'from watery scenes, fluid locations and aquatic environments' (p. 15). In reassessing how to engage in the task of thinking about water elsewhere that does not take devices for granted, Ballestero situates her work explicitly within the tradition of anthropology and acknowledges an affinity with feminist science studies and STS. 'This approach allows us to create an "opportunity to arouse a slightly different awareness" not only about "the problems and situations mobilising us" (Stengers 2005, p. 994), but also about the ways people confront those problems' (p. 29). Promoting a re-assessment of anthropologies engagement with water, Ballestero's ethnography of ordinary devices presents an alternative to thinking with water 'that invite[s] a renewed understanding of things we take for granted – a re-examination of our existing worlds and their political categories, through eyes open to wonder' (p. 35). Ballestero explains her method of how to conceptualise devices as ethnographic objects further:

> I want to suggest that wonder, that condition where it becomes imperative to think carefully about things that were presumed totally ordinary, and for that reason self-evident (Rubenstein 2006, p. 12), is a more generative disposition than crisis to think of how people … perform political work …Wonder opens up the familiarity of what seems straightforward.
>
> (p. 32)

Such positioning for Ballestero involves curiosity, doubt, and hesitancy on the part of the researcher. Employing wonder as a technique in her research and writing – 'as an epistemic mood' (p. 199) – from which to understand devices, Ballestero argues it is precisely the lack of wonder in research that so easily renders the device irrelevant and makes ordinary objects vulnerable to being ignored in our sense making. Throughout this book, I draw on Ballestero's definition of device as anthropological objects, and her conceptual and methodological tool of curiosity. The ethnographic focus on the mundane water fountain meets this definition of a device as an important participant in social life that has received little anthropological attention, whilst at the same time taking seriously curiosity as part of knowledge making processes with water: it offers curiosity as an entry point of how to understand the process of thinking with the water fountain, not just for the researcher but by people themselves in everyday sense making activities.

How to think, research, and write about water in STS

Some activists and scholars have pursued sense making activities with drinking water and called for new practices that engage consumers as participants, with Spackman (2023) launching initiatives like the Sensory Laboratory and bringing terms such as 'sensory politics' (Spackman and Burlingame 2018) to the forefront of discussion on water. In the article 'Sensory politics: The tug-of-war between potability and palatability in municipal water production', published in the special issue 'Toxic politics' in *Social Studies of Science*, Spackman and Burlingame (2018) discuss the politics and practices of sensory knowledge by publics and experts in creating and managing industrially produced tap water. However, as they point out in their contribution to the special issue, 'regimes of perception' about water quality in the 20th century marginalised consumers' sensory knowledge to the realm of the aesthetic, 'instead of accommodating it in the analytic' (p. 350). Sensory politics speaks to the practices of inclusion and exclusion of sensory knowledge in the making and regulation of municipal drinking water. Sensory politics shows how acknowledging water testing methods for making and regulating the quality of municipal water responds not just to the history of water as involving technological practices of standardisation and regulation, but also the classification and declassification of sensory knowledge. Sensory politics calls attention to the exclusion of everyday lay embodied experience from the classification of sensory knowledge, and the 'unevenness' (p. 367) of sensing methods. In showing how thinking with water involves embodied practices of taste and smell, Spackman and Burlingame highlight how the process of analysing water, performed by consumers in turning on the tap and drinking a glass of water, is codified as secondary. Consumers were conceptualised 'as aesthetically judging water, perhaps as connoisseurs, but never as analysts' (p. 363).

Importantly, Spackman's publications – 'Sensory politics' (with Burlingame) and *The taste of water* (2023) – do not ask us to merely recognise the unevenness of sensory labour and the marginalisation of consumer sensory practices in the making and regulation of drinking water. While, as Spackman observes, the practice of sensory labour belongs to professional water workers, the politics of sensory knowledge informs contemporary campaigns by municipal water companies to contest bottled plastic water. Using the example of the French 'Quvrez un grand cru' education campaign (2011–2015) initiated by Eau de Paris to educate all consumers of the quality of Paris tap water (discussed by Spackman 2023 in chapter 4), Spackman shows how municipal water producers appealed to consumers sensory knowledge to disrupt the binary of tap water/bottled water and dispute bottled waters' claims to taste. In training consumers to drink and evaluate water, Spackman's example illustrates how the public education campaign situates aesthetic practices of taste as embodied sites of public knowledge making. The example does not simply show us how consumers were invited to have a drink of water and ask questions in order to make situated aesthetic judgements on what they think of tap water; this example teaches us something about the way sensorial experiments on *how* to think with water concern the enrolment of publics as participants in the everyday contestation of plastic bottled water. More specifically, Spackman's example of municipal water producers mobilising a theatre of taste to create collective forms of public engagement – 'in a bottom-up form of sensory labour' – involve 'bring[ing] Paris's water back into the public sphere' (p. 149). Whilst peoples' response to the 'Quvrez un grand cru' campaign is mentioned briefly in Spackman's observations of its capacity to problematise bottled water through the apparatus of taste in local urban sites, the question of *how* this education campaign provoked thinking in public modes of participation is not the focus of her analysis.

This book attempts to address this process by situating London's #OneLess educational campaign as an empirical case study to investigate the contestation of plastic bottled water in everyday drinking practices. My empirical investigation of the public refill water fountain pilot draws on Veronica Strang's (2004, 2005) ethnographic study of domestic water use. Focusing on interactions with everyday water in specific cultural settings, Strang empirically investigates how people think and feel about water in ways that do not take technologies for granted. Using similar ethnographic methods to Strang, including fieldwork, close observation, and interviews, the aim of my study of public water fountains is to probe the aesthetic, sensorial and technical dimensions of everyday engagements with #OneLess' refill pilot in specific urban settings. Foregrounding the #OneLess refill experiment as an entry point to engage with how people think about drinking practices in situ, the focus of this book is to investigate the potential of mundane participatory practices to contest plastic bottled water.

In different ways, Whatmore and Landström's (2011, see also Whatmore 2013a,b) research on the experience of flooding in the North of England speaks to the importance of participatory methods as critical methodological tools for not only recognising the experience of lay publics and non-experts in research on environmental problems related to water, but the contribution of local participants to knowledge making. Noting that research in science studies has been based on 'observations of the activities of others' affected by the matter at issue, whilst the knowledge practice of researchers 'are not at stake in the controversy they are studying', Whatmore and Landström invent an apparatus – the competency group – for staging research collaborations between university experts and local members effected by the environmental knowledge controversy of flooding. The competency group 'complicates the "distance" between researcher and researched and the question of whose knowledge is at stake' (p. 604). Designed as a methodological experiment to overcome the distant relationship of experts to non-experts, the competency group involved the making and doing of a new research collaboration involving social scientists and natural scientists working collectively with local people affected by flooding. Putting Isabelle Stengers' (2005) experimental constructivism into practice, Whatmore and Landström's staging of the competency group deliberately posed challenges to the authority of professional experts from different disciplines. Professional knowledge claims and practices are 'put at risk/stake in the controversy alongside those of people affected by it with whom they collaborate' (p. 585). By creating a space in which expert knowledge claims 'could be more closely interrogated, and residents moved by the event of flooding to engage with the issues could try out alternative ways of framing and addressing the problem of flood-risk management informed by their experiences and observations' (p. 589), the aim of the competency groups was to slow down reasoning and open up questions about the production of knowledge and constitution of expertise. One of the primary techniques for achieving public engagement with the problem of flooding in the competency group was to work with 'brought objects (such as maps, photos, satellite images and even a remnant piece of mouldy carpet)' (p. 593) as artefacts involved in 'collective thinking' (p. 587). By inventing a research apparatus that slowed down expert reasoning and redistributed expertise, the competency groups working practice of ' "thinking with things" ' (p. 587) put at risk the knowledge practices of the group (including university and local members). It also brought to the fore the particular skills of social scientists 'in the design and conduct of research practices that stage more and different opportunities for new knowledge polities to emerge' (p. 606).

As Whatmore (2013a) demonstrates elsewhere, in 'Reflections on an experiment in geographical practice', published in the book *Interdisciplinarity: Reconfigurations of the social and natural sciences*, although geography as an interdiscipline is engaged in collaborations with natural scientists, the focus tends to be on environmental problem-solving. The potential of 'geography's interdisciplinary inventiveness' (p. 173) for '*re*constructing how environmental research is done' (p. 174, emphasis in original), Whatmore argues, is often left out of research aligned with governmental and business agendas. It was this gap that prompted the editors of this book Andrew Barry and Georgina Born (2013) to develop a framework for engaging with interdisciplinarity as a *problematisation*. Drawing on Foucault's notion of problematisation:

> 'the set of discursive or nondiscursive practices that makes something enter into the play of the true and false, and constitutes it as an object of thought (whether under the form of moral reflection, scientific knowledge, political analysis, etc.' (Foucault 2001, p. 1489; see also Rabinow 2002, p. 43). In order for this to occur, something 'must have happened to introduce uncertainty, a loss of familiarity; that loss, that uncertainty is the result of difficulties in our previous way of understanding' (Foucault 1994, p. 598),

Barry and Born (2013, p. 43) pose a positive conception of problems that directs us to the way the problematisation of certain situations 'may demand and generate novel responses'. Following Foucault,

Barry and Born understand interdisciplinarity as a means of generating questions, around which 'new forms of thought and experimental practice can coalesce' (p. 10). In their book, they point out how the focus of Whatmore's (2013a) flood risk project as a particular problematisation of the conduct of research, 'was not to address already defined problems, nor to engage in critique, but to generate positive problems through an intervention that responded to the need to invent novel practices and engagements' (p. 34). In involving non-experts in the practice of research, Whatmore's project 'both respond[s] to and lead[s] to the generation of new problems' (p. 44). *Problematising water* returns to Foucault's notion of problematisation, to take seriously these challenges of engaging experts and non-experts in the investigation and analysis of the #OneLess refill water fountain experiment. What might it mean to view the refill fountain as a provocation to think with water: to ask how are environmental campaigns to reduce plastic bottled water problematising, and respond not by engaging in critique, but inventing novel methods that involve non-experts and experts in the practice of research. The approach taken in this book responds to the interdisciplinary challenge of the refill project as a problematisation of the conduct of research.

As noted above, Whatmore and Landström's (2011) research apparatus of the competency group does not seek to merely critique environmental problems of flooding but study and model alternative interdisciplinary research practices. The staging of a competency group doubles as a performance of the making and doing of STS, one that aims to diversify the scope of the field and participate in the expansion of interdisciplinary research to engage both publics and professionals in forms of assessment on controversy. That is, Whatmore and Landström's experimental research technique has something to tell us in terms of how London's #OneLess refill water fountain experiment sparks connections that force thinking between the environmental problem of ocean plastic pollution and new forms of public drinking practices. As Noortje Marres and Javier Lezaun's (2011) opening to the special issue 'Materials and devices of the public: an introduction', which features Whatmore and Landström's article, put it, 'to understand how things acquire participatory capacities, we must attend to the empirical variability of materials and devices of the public' (p. 492). By considering publics as 'distinctively material productions', Whatmore and Landström's experimental competency group allows the 'material turn' to enter the study of specific political concepts, including participation (p. 499). In a long overdue consideration of technologies of participation, Marres and Lezaun assess Whatmore and Landström's conceptual and methodological intervention of staging the competency group in terms of its value to explore the concept of participation through participatory objects. Pondering 'how objects, devices, settings and materials acquire explicit political capacities, and how they serve to enact material participation as a specific public form', Marres and Lezaun (p. 489) perceive Whatmore and Landström's case study as offering a valuation of a set of artificial conditions intended to facilitate the production and observation of innovative forms of participation. Whatmore and Landström's contribution to the special issue is credited with helping to develop a research agenda in which science studies experiments become sites to consider 'their efficacy as multi-faceted devices of material participation' (p. 503).

We might look here at another text *Remaking participation: Science, environment and emergent publics* by Jason Chilvers and Matthew Kearnes (2015a), which refers to Whatmore and Landström's competency group as expanding the definition of collective participatory practice, to include 'material and non-human connections' (2015c, p. 270). Like Marres and Lezaun (2011), Chilvers and Kearnes (2015b) are interested in the science and technologies of participation. But they part ways to some extent with Marres and Lezaun when it comes to how we might empirically approach participation: their book theorises the specific ways STS scholars and critical social scientists have 'taken formalised techniques, tools, and instruments for public involvement as their main entry point' (p. 42). The technologies of participation approach, they argue, has a tendency to foreground the role of social science methods in enacting 'mini-publics' (p. 46), whilst underplaying the role of emotions and importance of embodied and sensory elements in the 'emergence of participatory experiments' (p. 40). This move, according to Chilvers and Kearnes, carries the 'risk of reifying the particular technologies of participation at the expense of other possibilities and imaginations of participatory politics' (p. 46). By turning their attention from the study of scientific practices to social scientific and participatory ways of knowing, representing, and mobilising publics, Chilvers and Kearnes develop an alternative analytic interpretive agenda for social studies of participation. Their focus addresses the co-productive, relational, and emergent nature of the process of collective sense making in specific situated participatory experiments and practices. Viewing practices of public participation and engagement as a 'contingent and heterogeneous collective of human and non-human actors, devices, settings, theories, social science methods, public participation, procedures and other artefacts' at particular sites, they ask:

- How do participatory experiments and practices come into being, what shapes them and how are they articulated?
- Who/what enrols heterogenous actors into participatory collectives, mediates them and with what significant exclusions?
- What resistances and challenges do participatory experiments face?
- What are the productive dimensions and effects of collective participatory practices in terms of the ways in which they construct the issues or objects of participation, identities of the actors involved, material commitments and models of participation of democracy itself?

(pp. 33–34)

By addressing the practical challenge of cultivating a co-productionist and relational way of opening participation practices to empirical analysis, Chilvers and Kearnes' (2015c) questions move the focus of inquiry from prescribing in advance what participation is and offer instead answers on how to engage with these questions in practice. Crafting a methodology that moves towards understanding the diversities of participatory practices for Chilvers and Kearnes entails; 'defining and mapping out the space of coherence in question as a starting point', through the use of issue mapping techniques and assistance of more traditional research methods such as multi sited ethnography (Marcus 1995), documentary evidence, and ethnographic interviews (pp. 273–274); engaging with mediators and experts of participation 'to prompt reflexive learning' (p. 276); taking seriously 'the emergence of public collectives', and the ways in which public meanings are constructed and represented (p. 280); and a 'commitment to nurturing relational-reflexivities in a more thoroughly systemic and distributed sense' that reframes them from the perspective of all actors (p. 281). This book takes up Chilvers and Kearnes' (2015b,c,d) methodology for investigating collective participatory practices in situated spaces of participation in relational co-productionist terms. In my attempt to make explicit and broaden the repertoire of empirical research on water to include the refill water fountain as a starting point from which to conduct an investigation of the #OneLess education campaign, I engage in the use of research methods of ethnographic interviews, multi-sited ethnography, and close observation to investigate the mundane technology of the water fountain, as well as the publics, experts, mediators, and everyday spaces of experimentation in relational reflexive knowledge practices. But in approaching the refill water fountain pilot empirically, with the intent of determining whether or not the #OneLess experiment succeeds or not in contesting plastic bottled water and reducing plastic ocean pollution, Chilvers and Kearnes' (2015b,c,d) conceptual and methodological lens for engaging participatory practices, collectives, and experiments risks replicating the prescriptive and reifying tendencies of the technologies of participation paradigm in critical social science methods and STS, and detracts from the task of empirical inquiry: to investigate *how* the #OneLess campaign incites questions and problems rather than merely resolving them. While Chilvers and Kearnes' (2015b,c,d) social studies of participation paradigm offers important insights into the collective, relational, and emergent nature of participation in the #OneLess experiment, this book suggests we cannot dispense with Foucault's notion of problematisation for engaging the process of how thinking, research, and writing with the refill water fountain happens in practice.

A problematological turn

In the introduction to the special issue 'Problematizing the problematic' in the journal *Theory, Culture & Society*, Martin Savransky (2021) makes the academic move of seeking 'to contest, trouble and problematise the conception of problems' (p. 5). Noting that 'there is no practice of problematisation that does not involve a problematisation of our own practices', Savransky (p. 15) zeroes in on the particular use of Foucault's notion of problematisation in the work of Stengers (2021) and Barry (2021) in their contributions to the special issue. Referring to problematisation, not as mere analytical exercise but an '*art of metamorphosis*' 'of our modes of knowing, of living, of being', Savransky (2021, p. 20, emphasis in original) points out that the aim of these authors is not to engage in a debate about the best way to read Foucault's concept of problematisation. The articles, for Savransky, are more concerned with the implications for social theorists and social scientists that follow from using problematisation in empirical studies. In Stengers' invitation to think about how Foucault's understanding of problematisation shapes research and thinking on the climate crisis in practice, she points to a double move in Foucault's (1984a,b) work on problematisation. On the one hand, a 'mode of historical inquiry into practices that render something as an object of thought (as moral reflection, scientific knowledge, political concern)' (p. 76), and on the other, a philosophical ethos, involving 'a *transformative* engagement, an engagement which forces the thinker to test the limits of thinking' (p. 76, emphasis in original). The study of modes of problematisation, Stengers points out, is a situated one. It is marked 'by a practical situation becoming a matter of explicit concern' that 'cannot be met by epistemology or by the acquisition of (critical or positive) knowledge'. But rather involves a questioning situation that forces 'one to "problematize", to put ones critical work to the test of experimentation with other ways of living and of thinking' (p. 79). Stengers' insights prompt Savransky to argue 'problematisation, in the end, in the beginning, is always *an art of metamorphosis*' (p. 15, emphasis in original).

With this definition in hand, we can see environmental problems on the whole as problematising, as does Barry (2021) in his article 'What is an environmental problem?'. Here, Barry refers to the challenges posed by Foucault's (1989) notion of problematisation to Actor Network Theory (ANT) and urges STS scholars and social scientists to extend the empirical programme of research on environmental problems to 'the study of how problems are generated and transformed' (p. 111). This includes, he argues, being alert to the 'ongoing transformations of living environments that participate in the generation of problems' (p. 112). For Barry, developing a critical empirical orientation of engagement with what a local environmental problem is in particular situations involves; 'following the abductions of participants' (p. 112), as well as the 'claims of those who know in advance what the solution to the problem is and of those who know what hidden forces have determined what has happened' (p. 113). This dual process is understood by Barry as demanding rigorous empirical sociological research, insofar as it:

> should direct the researcher to interrogate the boundaries between what can and cannot be problematised, what can and cannot be rendered public, and what voices and arguments have and have not been heard – including those responses and feelings that may 'exist

> beyond the threshold of conscious communication and intent' (Colebrook 2008, p. 127).
>
> (p. 112)

Following Barry, in the introduction to the special issue Savransky argues an empirical study of the formation of environmental problems can only be articulated through a practice of '*abductive problematisation*: risking a leap beyond what is given in perception so as to trouble the publicly established boundaries through which a problem is posed, precipitating an amplification of tensions while resisting the resolution of a situation by means of analytical integration' (p.15, emphasis in original). Savransky's description of Barry's sociological insights prompt us to look at the uses of problematisation in research, noting that research on environmental problems has been locked into a theoretical approach to problematisation at the expense of more thorough empiricist approaches. While Barry's preference for abductive problematisation aims to expand the links between Foucault's description of problematisation and the task of showing how research may 'generate multiple solutions, and indeed, further problems' (p. 95), Barry does not expand on the empirical tools required for investigating these questions. This book attempts to make some of these additional strides into problematisation as less a theoretical than a methodological problem for researchers. To do so, I turn to the late Clive Barnett's (2015) (see also Barnett and Bridge 2017) and Thomas Osborne's (2003) (see also Osborne and Rose 1997, 2004; Osborne, Rose and Savage 2008) engagements with Foucault's provocation of problematisation as an opportunity to help us think beyond narrow definitions of the concept and engage in the task of developing novel methods that can be used to address everyday modes of problematisation in the specific settings of the #OneLess refill fountain experiment.

In almost two decades before Barry posed the question 'what is an environmental problem?', Osborne (2003) in his article 'What is a problem?' sought to replace the emphasis on the diagnosing of problems or dictating of solutions with a methodology 'anthropologically "committed" to problematology' (p. 2), in which problems are prioritised over theories. By taking Foucault's (1984a) problematic of 'analysing – and thus specifying, isolating – this level of problematics and problematisations itself' (p. 11) as a starting point for research, Osborne's methodology of problematology seeks to diversify the scope of engaging problematisations as creative events.

Drawing on Foucault's (1988) work on problematisation, Osborne and Rose (1997) identify a major gap in the voicing of problems in social thought. In their article 'In the name of society, or three theses on the history of social thought', they find that while a social perspective on problems is assumed to be the provenance of social philosophers and sociologists, privileging the voice of the academic author as unique in the voicing of society, they argue, sidesteps and marginalises non-academic figures that bring into existence new ways of problematising. Adapting Gilles Deleuze and Felix Guattari's (1994) '*conceptual persona*' to describe non-academic inventive figures as thinkers who 'embody a particular style or stance towards a set of problems, articulating them in a particular orientation' (p. 94, emphasis in original), Osborne and Rose point out the most important conceptual personae in the 19th and 20th century were not theorists, but practical people like doctors, engineers, and reformers. In staying close to a particular zone of problematisation, these practical thinkers 'articulate a novel way of answering the questions it poses' (p. 95). This mode of thinking problematically depended on novel ways of construing practical problems in terms of 'a space of action' that exceeded the individual (p. 89) and included flows of water. By engaging with particular problems in everyday spaces and devising empirical contraptions 'to realise possible solutions to them' (p. 94), Osborne and Rose go on to point out that the earliest social thinkers, as Foucault himself recognised, ' "were specialists of space" (1980, p. 151)' (2004, p. 225). In seeking to address how certain things become articulated as problems and the ways this happens in spaces of action, Osborne and Rose (1997) look to social research techniques such as the research interview and observation to 'summon up in reality phenomena that have already been summoned up in thought' (p. 100). What counts in the elaboration of a specific practical problematisation, they argue, are the technical procedures 'that make things "visible" ' (p. 98).

In the decades since Osborne and Rose made their proposal on the inventiveness of social thought by conceptual persona in specific problem spaces, they have reflected further on Foucault's notion of problematisation and the methodological challenges this poses for the discipline of sociology and its methods:

> To analyse the history of such question-fields is to investigate why certain things (behaviour, phenomena, processes) become articulated *as* questions, how they are linked up with or divided

> off from other phenomena, and the various ways (conditions and procedures) in which this actually happens. Social thinkers, as we well know, tend to operate in the problem spaces that are created for them by others … When they are inventive, *their work transforms those problem spaces, makes them thinkable in new ways*. But, we suggest, this inventiveness is not usually related to the individual genius of individuals, but rather to the *techniques through which problems and questions about the social world are made visible.* Hence, what is important is not the originality of this or that great thinker but the techniques that are available to thought at any particular time and place.
>
> (Osborne, Rose and Savage 2008, p. 526, my emphases)

In order to address the problems of our social world and the techniques through which they become thinkable, Osborne, Rose, and Savage (2008) call for a descriptive form of sociology. Descriptive investigations that seek to render visible how certain things become articulated as questions and problems involve conceptualising their connectedness to problem spaces and empirically tracing the connections between things. The potential of such a descriptive turn is seen to enhance opportunities for staging collaborations with natural scientists and non-academic experts. This book seeks to extend such descriptive endeavours (see Vitellone, Mair and Kierans 2021).

In looking to investigate how certain things become articulated as questions and problems, and the spaces and techniques through which they become visible and thinkable, this book aligns itself with Foucault's task of problematisation, as a matter to direct our curiosity to the ways the #OneLess refill pilot gets linked up with the contestation of bottled plastic water, and the way this actually happens in practice. Analysing the space of the water fountain from the point of view of problematisation will be situated and employed to engage the different humans and non-humans, conceptual personae, experts, and publics involved in the #OneLess refill experiment, offering an interdisciplinary approach to environmental problems that cannot be known, nor solved, with the tools from individual and siloed disciplines, and how problematisation might be investigated descriptively to enliven research, thinking, and writing with water. Chapters 1 and 2 will spend some time introducing the theoretical and methodological turn to problematisation taken in this book and how thinking problematically with the water fountain matters. Using the water fountain as a case study for reflection and insight, these chapters attend to the role of problematisation in existing studies and offer an alternative framework to address the plastics and climate crises, showing what a descriptive mode of inquiry provides for challenging and transforming the task of engaging with the environmental problem of plastic pollution and what this means for practitioners, publics, and researchers. It does so by employing, probing, and further developing Foucault's notion of problematisation as a concept for inquiring into the mundane practice of drinking water in urban public settings. Outlining the books engagement with Foucault's notion of problematisation in relation to my case study of the water fountain, as a phenomena in the world to be studied and engaged with empirically, will be developed via an engagement with relevant literature in the fields of human geography (Chapter 1), sociology (Chapters 2 and 3), interdisciplinary studies (Chapter 3), anthropology (Chapter 4), STS (Chapters 2, 3, and 4), and ethnomethodology (Chapter 5) and employed as a methodological approach for studying the #OneLess refill water fountain pilot in Chapters 3, 4, and 5. The empirical chapters (3, 4, and 5) will explore this trajectory by introducing the #OneLess refill experiment led by the Marine and Freshwater Team in the Conservation and Policy Department of ZSL, as an empirical case study to move the focus of inquiry from academic critiques of the global environmental problem of plastics pollution and access to free drinking water towards an investigation of problem spaces and situated practices of problematisation.

Content of the book

The chapters in *Problematising water: An experiment on the edges of interdisciplinarity* are a manifestation of thinking with the water fountain across a range of disciplines including human geography, sociology, anthropology, STS, and ethnomethodology. The chapters are structured around questioning divergent methodological, epistemological, and ontological assumptions and probing new ways of addressing the water fountain including ethnographic methods to investigate #OneLess' refill experiment as a particular problematisation. Chapter 1 summarises the challenges of engaging the water fountain as an object of inquiry and imagines a new sense of purpose for Foucault's concept of problematisation. In order to understand what is at stake in using Foucault's methods for water-focused research we contextualise problematisation in developments in the field of human geography that emerged in the 2010s. Focusing on Linton's (2010) Foucauldian-inspired history of water, we consider the example of the water fountain in Linton's knowledge practices and its impact for stimulating theorising, research, and writing with water in the discipline of human geography. Comparing the reaction of the critical human geographers Ekers and Loftus (2008) to Linton's hydrosocial research paradigm with Barnett's (2015) and Barnett and Bridge's (2017) approach to thinking problematically as an urban practice, the chapter opens up another way of responding to Linton's study of water that enables a re-imagining of hydrosocial knowledge practices and the material means by which to make sense of plastic bottled water and its contestation.

Chapter 2 takes on the theoretical and methodological challenges of engaging with the water fountain in the social science disciplines of sociology and STS. To do so, it addresses Hawkins, Potter, and Race's (2015) study of mobile drinking practices and the 'Do Something' awareness campaign to re-introduce public water fountains in Sydney, Australia. Focusing on Hawkins et al.'s case study of plastic water and academic engagements with it, the chapter draws attention to two contrasting uses of problematisation in research practice. On the one hand, problematisation as a method of analysis, on the other, problematisation as a phenomenon of how things become a problem and matter of concern for people. Inspired by Stengers (2021), the chapter takes seriously the task of testing what problematisations do in the context of the climate and plastics crises. Problematisation, for Stengers, involves engaging situated practices of thinking on which live problematisations are formed. Investigating problematisation in a collective shared sense around a problematic situation, actualised in concrete apparatuses that gather concerned people, requires relearning how thinking and knowledge becomes cultivated in everyday public situations and situated practices. By probing problematisation, not as a method of critical analysis, but a transformative everyday practice, the chapter outlines diverse ways of theorising the climate and plastics crises and proposes a new way to address everyday problematisations in environmental experiments. Using the water fountain as a case study to show how Foucault's concept of problematisation can contribute to the further development of social scientists' capacity to think problematically, the chapter opens new methodological insights on how to conduct research on the water fountain.

Chapter 3 takes a step into my research on the #OneLess refill water fountain pilot. Here we address the role of interviewing and device of the audio recorder as enabling an interdisciplinary collaboration with the #OneLess project team. To do so, the chapter draws together literature analysing interdisciplinarity and problematisation in relation to environmental studies (Barry and Born 2013), with literature analysing the audio recording device in the social sciences (Back 2012; Lury and Wakeford 2012; Michael 2012, 2022). In combining these insights, the chapter goes on to illustrate how the audio recorder became central to developing a collaborative interdisciplinary research project for researching #OneLess' refill pilot that enabled an encounter between natural and social scientists, publics, and professionals. Reflecting on the device of my audio recorder in the conduct of research, the chapter proceeds to illustrate how the method of interviewing enabled me to question disciplinary modes of thinking and probe new ways of evaluating the impact of #OneLess' refill water fountain experiment.

Chapter 4 examines the public existence of the global environmental problem of ocean plastic pollution in specific local public settings of #OneLess' refill water fountain pilot. In this chapter, and the following chapter, we address the impact and effects of London's refill experiment on the public. Drawing on my fieldnotes taken from participant observation across three sites during different periods of fieldwork (2018–2019), the chapter discusses the consequences of using ethnographic methods of close observation and thick description to connect to the way the environmental problem of plastic ocean pollution involves everyday modes of problematisation in the urban setting of the refill fountain. Aligning descriptive practices with the presence of humans and technical objects in the mundane settings of the refill pilot, the chapter discusses the contribution of descriptive methods for investigating everyday problematisations from the point of view of their achievement and illustrates the use of sociography for evaluating the #OneLess' campaign to change Londoners' drinking practices.

Chapter 5 continues to engage the problem space of the London's #OneLess refill pilot. Here we address how plastic bottled water becomes differently articulated as a problem in the voice of practitioners and publics. Drawing on qualitative data gathered from 4 interviews with the marine scientists and conservationists and systems change practitioners responsible for the design and implementation of

#OneLess' campaign, and 130 interviews with refill water fountain users in London over a 6-month period (2018–2019), the chapter compares the mapping methods of systems change practitioners with the practical methods and concrete situations around which people gather and make sense of their worlds in common. In giving a voice to how problematisations are differently articulated in thought by practitioners and research participants engaged with the water fountain, we take seriously the situated activity of refilling as an accountable and thus recountable inter/action (Garfinkel 1967/1984, 2002). Focusing on the inter/action between the refill water fountain and the mundane work of refilling as a methodic apparatus to render social phenomena explorable and analysable, we see what studies in ethnomethodology contribute to understanding the impact of the #OneLess refill experiment and the role of Foucault's notion of problematisation as a resource to articulate how thinking with the water fountain happens. By making visible how problematisation is articulated as a practical problem 'in and as of' the everyday settings of the refill fountain (Garfinkel 2002, p. 92), the chapter shows how ethnomethodological practice and writing shape data in ways that allow for the making, interpretation, and translation of findings as an interdisciplinary problematisation.

References

Anand, N. (2017) *Hydraulic city: Water and the infrastructures of citizenship in Mumbai.* Duke University Press.

Back, L. (2012) Tape recorder. In C. Lury and N. Wakeford [Eds.] *Inventive methods: The happening of the social* (pp. 245–260). Routledge.

Bakker, K. (2003) *An uncooperative commodity: Privatising water in England and Wales.* Oxford University Press.

Ballestero, A. (2019) *A future history of water.* Duke University Press.

Barad, K. (2006) *Meeting the universe halfway: Quantum physics and the entanglement of matter and meaning.* Duke University Press.

Barnes, J. and Alatout, S. (2012) Water worlds: Introduction to the special issue of Social Studies of Science. *Social Studies of Science* 42(4), 483–488.

Barnett, C. (2015) On problematization: Elaborations on a theme in 'Late Foucault'. *Nonsite.* Issue 16, June 22.

Barnett, C. and Bridge, G. (2017) The situations of urban inquiry: Thinking problematically about the city. *International Journal of Urban and Regional Research* 40(6), 1186–1204.

Barry, A. (2021) What is an environmental problem? *Theory, Culture & Society* 38(2), 93–117.

Barry, A. and Born, G. (2013) Interdisciplinarity: Reconfigurations of the social and natural sciences. In A. Barry and G. Born [Eds.] *Interdisciplinarity: Reconfigurations of the social and natural sciences* (pp. 1–56). Routledge.

Bijker, W. (2007) Dikes and dams, thick with politics. *Isis* 98(1), 109–123.

Chen, C., MacLeod, J. and Neimanis, A. (2013) Introduction: Toward a hydrological turn? In C. Chen, J. MacLeod and A. Neimanis [Eds.] *Thinking with water* (pp. 3–22). McGill-Queens University Press.

Chilvers, J. and Kearnes, M. [Eds.] (2015a) *Remaking participation: Science, environment and emergent publics*. Routledge.

Chilvers, J. and Kearnes, M. (2015b) Science, democracy and emergent publics. In J. Chilvers and M. Kearnes [Eds.] *Remaking participation: Science, environment and emergent publics* (pp. 1–27). Routledge.

Chilvers, J. and Kearnes, M. (2015c) Participation in the making: Rethinking public engagement in co-productionist terms. In J. Chilvers and M. Kearnes [Eds.] *Remaking participation: Science, environment and emergent publics* (pp. 31–63). Routledge.

Chilvers, J. and Kearnes, M. (2015d) Remaking participation: Towards reflexive engagement. In J. Chilvers and M. Kearnes [Eds.] *Remaking participation: Science, environment and emergent publics* (pp. 261–288). Routledge.

Colebrook, C. (2008) Bourgeois thermodynamics. In I. Buchanan and N. Thoburn [Eds.] *Deleuze and politics* (pp. 121–138). Edinburgh University Press.

de Laet, M. and Mol, A. (2000) The Zimbabwean bush pump: Mechanics of a fluid technology. *Social Studies of Science* 30(2), 225–263.

Deleuze, G. and Guattari, F. (1994) *What is philosophy?* Trans G. Burchell and H. Tomlinson. Verso.

Ekers, M. and Loftus, A. (2008) The power of water: Developing dialogues between Foucault and Gramsci. *Environment and Planning D: Society and Space* 26(4), 698–718.

Foucault, M. (1984a) Polemics, politics and problematizations. In P. Rabinow [Ed.] *The Foucault reader* (pp. 381–390). Pantheon Books.

Foucault, M. (1984b) What is enlightenment? In P. Rabinow [Ed.] *The Foucault reader* (pp. 32–50). Pantheon Books.

Foucault, M. (1988) On problematization. *History of the Present* 4, 16–17.

Foucault, M. (1989) The concern for truth. An interview with Francois Ewald. In S. Lotringer [Ed.] *Foucault live* (pp. 293–308). Semiotext(e).

Foucault, M. (1994) Polemics, politics and problematisations. In M. Foucault, *Ethics: The essential works, Volume 1* (pp. 111–120). Penguin.

Foucault, M. (2001) Le souci de la verite. In M. Foucault, *Dits et ecrits II, 1976–1988* (pp. 1487–1497). Gallimard.

Garfinkel, H. (1967/1984) *Studies in ethnomethodology*. Polity.

Garfinkel, H. (2002) *Ethnomethodology's program: Working out Durkheim's aphorism*. Edited with an Introduction by A.W. Rawls. Rowman and Littlefield.

Gentilcore, D. (2023) In praise of the ordinary: Shifting knowledge and practice in the medical use of drinking water in Italy, 1550–1750. *Bulletin of the History of Medicine* 97(4), 531–559.

Hamlin, C. (1992) *A science of impurity: Water analysis in 19th Century Britain.* University of California Press.

Hamlin, C. (2000) 'Waters' of 'water'? – master narratives in water history and their implications for contemporary water policy. *Water Policy* 2(4–5), 313–325.

Hastrub, K. and Hastrub, F. [Eds.] (2015) *Waterworlds: Anthropology in fluid environments*. Bergham.

Hawkins, G. (2011) Packaging water: Plastic bottles as market and public devices. *Economy and Society* 40(4), 534–552.

Hawkins, G. (2013) The performativity of food packaging: Market devices, waste crisis and recycling. *The Sociological Review* 69(S2), 66–83.

Hawkins, G., Potter, E. and Race, K. (2015) *Plastic water: The social and material life of bottled water*. MIT Press.

Jensen, C. (2017) Pipe dreams: Sewage infrastructure and activity trails in Phnom Penh. *Ethnos* 82(4), 627–647.

Kaika, M. (2005) *City of flows: Modernity, nature and the city*. Routledge.

Kaplan, M. (2011) Lonely drinking fountains and comforting coolers: Paradoxes of water value and ironies on water use. *Cultural Anthropology* 26(4), 514–541.

Kirby, V. (2008) Natural convers(at)ions: Or, what if culture was really nature all along? In S. Alaimo and S. Heckman [Eds.] *Material feminisms* (pp. 214–237). Indiana University Press.

Lavau, S. (2013) Going with the flow: Sustainable water management as ontological cleaving. *Environment and Planning D: Society and Space* 31 (3), 416–433.

Liboiron, M. (2021) *Pollution is colonialism*. Duke University Press.

Liboiron, M. and Lepawsky, J. (2022) *Discard studies: Wasting systems and power*. MIT Press.

Linton, J. (2010) *What is water?: The history of a modern abstraction.* University of British Columbia Press.

Linton, J. and Budds, J. (2014) The hydrosocial cycle: Defining and mobilising a relational-dialectical approach to water. *Geoforum* 57, 170–180.

Lora-Wainwright, A. (2017) *Resigned activism: Living with pollution in rural China*. MIT Press.

Lury, C. and Wakeford, N. (2012) Introduction: A perpetual inventory. In C. Lury and N. Wakeford [Eds.] *Inventive methods: The happening of the social* (pp. 1–24). Routledge.

Mah, A. (2022) *Plastic unlimited: How corporations are fuelling the ecological crisis and what we can do about it*. Polity Press.

Mah, A. (2023) *Petrochemical planet: Mulitscalar battles of industrial transformation*. Duke University Press.

Marcus, G.E. (1995) Ethnography in/of the world system: The emergence of multi-sited ethnography. *Annual Review of Anthropology* 24, 95–117.

Marres, N. and Lezaun, J. (2011) Materials and devices of the public: An introduction. *Economy and Society* 40(4), 489–509.

Michael, M. (2012) De-signing the object of sociology: Toward an idiotic methodology. *The Sociological Review* 60(S1), 166–183.

Michael, M. (2022) *The research event: Towards prospective methodologies in sociology*. Routledge.

Neimanis, A. (2014) Alongside the right to water, a posthumanist feminist imaginary. *Journal of Human Rights and the Environment* 5(1), 5–24.

Neimanis, A. (2019) *Bodies of water: Posthumanist feminist phenomenology*. Bloomsbury.

Osborne, T. (2003) What is a problem? *History of the Human Sciences* 16(4), 1–17.

Osborne, T. and Rose, N. (1997) In the name of society, or three theses on the history of social thought. *History of the Human Sciences* 10(3), 87–104.

Osborne, T. and Rose, N. (2004) Spatial phenomenotechnics: Making space with Charles Booth and Patrick Geddes. *Environment and Planning D: Society and Space* 22(2), 209–228.

Osborne, T., Rose, N. and Savage, M. (2008) Reinscribing British sociology: Some critical reflections. *The Sociological Review* 56(4), 519–534.

Pia, A. (2024) *Cutting the mass line: Water, politics and climate in Southwest China*. John Hopkins University Press.

Rabinow, P. (2002) Midst anthropology's problems. *Cultural Anthropology* 17(2), 135–149.

Rubinstein, M.J. (2006) A certain disavowal: The pathos and politics of wonder. *Princeton Theological Review* 12(2), 11–18.

Savransky, M. (2021) Problems all the way down. *Theory, Culture & Society* 38(2), 3–23.

Sheard, S. (1994) Water and health: The formation and exploitation of the relationship in Liverpool, 1847–1900. *The Historic Society of Lancashire & Cheshire* 143, 141–163.

Sofoulis, Z. (2005) Big water, everyday water: A sociotechnical perspective. *Continuum: Journal of Media and Cultural Studies* 19(4), 445–463.

Spackman, C. (2023) *The taste of water: Sensory perception and the making of an industrialized beverage*. University of California Press.

Spackman, C. and Burlingame, G. (2018) Sensory politics: The tug-of-war between potability and palatability in municipal water production. *Social Studies of Science* 48(3), 350–371.

Stengers, I. (2005) The cosmopolitan proposal. In B. Latour and P. Weibel [Eds.] *Making things publics: Atmospheres of democracy* (pp. 994–1003). MIT Press.

Stengers, I. (2021) Putting problematization to the test of our present. *Theory, Culture & Society* 38(2), 71–92.

Strang, V. (2004) *The meaning of water*. Berg.

Strang, V. (2005) Common senses: Water, sensory experience and the generation of meaning. *Journal of Material Culture* 10(1), 92–120.

Swyngedouw, E. (2004) *Social power and the urbanisation of water: Flows of power*. Oxford University Press.

Taylor, V. and Trentmann, F. (2011) Liquid politics: Water and the politics of everyday life in the modern city. *Past and Present* 211(May), 199–241.

Vitellone, N., Mair, M. and Kierans, C. (2021) Doing things with description: Practices, politics, and the art of attentiveness. *Qualitative Research* 21(3), 313–323.

Whatmore, S. (2013a) Where natural and social science meet? Reflections on an experiment in geographical practice. In A. Barry and G. Born [Eds.] (2013) *Interdisciplinarity: Reconfigurations of the social and natural sciences* (pp. 161–177). Routledge.

Whatmore, S. (2013b) Earthly powers and affective environments: An ontological politics of flood risk. *Theory, Culture & Society* 30(7/8), 33–50.

Whatmore, S. and Landström, C. (2011) Flood apprentices: An exercise in making things public. *Economy and Society* 40(4), 582–610.

1 Thinking problematically with the water fountain

Jamie Linton's (2010) book *What is water?* produced new ways of helping us make sense of water that insists answering this question is difficult without involving people and technology. Focusing on Linton's hydrosocial research and how it has been theorised by scholars in the discipline of human geography, this chapter addresses developments in the discipline and offers new modes of thinking and ways of engaging with the water fountain. For Linton, water 'always reflects, in one way or another the state of society' (preface, p. 1). Whilst Linton's approach has been acknowledged and debated for bringing together Marxist dialectical thinking and Bruno Latour's Actor Network Theory (ANT) and concept of hybridity, the influence of Michel Foucault's thinking for stimulating Linton's hydrosocial approach is less developed. Addressing what we can learn from Linton's Foucauldian-inspired genealogy of water, this chapter offers an interpretation of Linton's question what is water as a problematisation and provides a conceptual and methodological approach for analysing it. In situating and unfolding Foucault's (1984) notion of problematisations in the practice of theorising water, the chapter takes Linton's example of the water fountain as a starting point to challenge and transform scholarly practice. By making explicit the example of the water fountain as an object to examine problematisation in practice, the chapter moves the focus of inquiry in Linton's hydrosocial paradigm from critical applications of theory to employing problematisation as a concept that can be studied and engaged with empirically.

DOI: 10.4324/9781003616757-2

Foucault meets hydroscience: developing a method of analytical and practical hydrolectics

Drawing on Foucault's methods, Linton's book *What is water?* traces the way thinking about water in the 20th century has shifted. In his book, Linton identifies historical changes in ways of engaging with water from a material abundance that we took for granted to something we can no longer ignore and are more likely to associate with scarcity, pollution, and crisis. Reflecting on the predominate way of thinking about water in the historical present – in relation to concerns around the global water crisis, water scarcity and water pollution, uneven access to water, and the consequences of climate change – Linton argues, we have 'begun to *think* about water in a different way' (p. 6, emphasis in original). The ecological, cultural, and political aspects of water, he points out, are presented to us 'in ways that challenge and defy our abstract understandings of water's nature' (p. 7). What concerns Linton is the articulation of problems in terms of a global water crisis and the framing of solutions in technical and hydrological terms 'diverts attention from the political and social circumstances that produce such problems' (p. 8). Ways of imagining, representing, and materialising modern water at any historical time and place, argues Linton, reflect the 'relative power of different social actors' to fix the meaning of water (p. 13), making less real and visible 'alternative kinds of water' (p. 11). What's required to challenge the hegemonic construction of water, according to Linton, is an alternative approach that 'deliberately recognises, or restores, its social nature' (p. 23). Overcoming the hegemony of modern water involves 'changes in how we think about water, as well as how we represent, manage, distribute, value and use it, for all these are closely related' (p. 13).

In order to challenge the concept of modern water as something abstract to be governed, to something 'more complicated' (p. 6), Linton provides a hydrosocial conceptual framework of waters as diverse, that are 'made and are made known through different modes of social engagement' (p. 228). Conceptualising and investigating water involves a mix of social and hydrological processes that cannot be isolated from each other. In order to understand the relationship between water and society, Linton (see also Linton and Budds 2014) advances the concept of the hydrosocial cycle as a means of theorising and analysing water–society relations. The hydrosocial cycle is proposed as an analytical tool for geographers and social scientists to produce practical and 'critical knowledge of the social nature of waters' (p. 231). Applying Foucault's genealogical approach for studying the history of things to the study of water, Linton proposes a program of hydrosocial research that does not take for granted the 'way of knowing water and the habits of thought and practice that it helps induce' (p. 8). What this entails in practice involves developing new ways of seeing water, what it is, and how it is known in relation to the social that are no less real or legitimate than dominant ways of knowing modern water as a fixed, natural, and universal phenomena. Outlining how Foucault's methods align with his own, Linton expands on the connection between the concept of hydrosocial cycle and a genealogical investigation of water:

> what we now presume to be the basic nature of water is actually a remarkable accomplishment, one that can be traced through the history of relations between water and people, including the myriad uses to which water has been put, the attachments of various meanings to water, the social differences and conflicts that have been mediated by water, and the ways that water has been made and know to (and by) philosophers and scientists.
>
> (p. 9)

Attending to the social circumstances that make water what it is in Linton's hydroscientific program of research has two key strands. On the one hand, reconstituting the meaning of water in relation to the social involves a method of analytical hydrolectics. Analytical hydrolectics requires attending to the ways '*waters* are diverse, are made and made known through different modes of social engagement' (p. 228, emphasis in original), including mundane processes such as drinking water. The aim of analytical hydroelectrics is 'to gain a comprehensive appreciation of the full range of actors and waters involved with respect to the problem or issue in question' (p. 246) including stakeholders, participants, and users involved in decisions, problems, and issues. In order to demonstrate the socio-historical nature of water in specific contexts, Linton uses his own personal experience of using a water fountain at Queens University campus in Kingston, Ontario. Linton's case study of the water fountain is used to make visible the hydrosocial cycle as a process for analysing water–society relations and illustrate his method of analytical hydrolectics. Linton's practice of analytical hydrolectics involves tracing and narrating '*how people relate to water*' (p. 211, emphasis in original), including visual methods of drawing the hydrosocial cycle through photographic devices of image making. Using his own photographs of public water fountains in Canada, Linton's book provides three figures to narrate the 'performance' of the hydrosocial cycle as a means of producing knowledge of the social nature of waters; 'Figure 12.2 the cycle of public water and the production of a body politic: a drinking fountain', 'Figure 12.3 Interruption of the cycle of public water: a drinking fountain blocked by a bottled water vending machine', and 'Figure 12.4 Diversion from the public water cycle to a private vending machine and the production of individual consumers' (pp. 231–234). In these examples, Linton compares how the use of a public drinking fountain and the bottled water vending machine affect social relations. These three figures are used as empirical sources to 'illustrate two different *kinds* of waters, as defined by the social structures and physical infrastructures by which they are made available to people' (p. 231, emphasis in original).

In a follow up article to *What is water?* Linton and Jessica Budds (2014) make further reference to Linton's example of the public water fountain as an illustration of the hydrosocial cycle. In doing so, their goal is to demonstrate how specific technologies for providing drinking water 'are internally related to *particular kinds of waters and particular kinds of social relations*' (p. 175, emphasis in original). By using the water fountain as a key example to make visible the hidden process of altering water flows (through interrupting and blocking access to public drinking water), Linton's methodological approach for engaging the hydrosocial cycle and how it 'becomes produced in, and how it configures social relations' (p. 177) does not take for granted the specific technologies in 'any given situation' (p. 178). For Budds, Linton, and Rachael McDonnell (2014) the mundane technology of the water fountain exemplifies the process of the co-constitution of water and society and illustrates an analytical framework of hydrolectics, 'by reflecting upon what water is' and 'how water is known' (p. 168).

Addressing the question *What is water?* in the hydrosocial cycle also involves a method Linton refers to as practical hydrolectics. In contrast to the approach of analytical hydrolectics, which involves the academic interrogating water and analysing social relations through methods of illustration and narration, practical hydrolectics reorients the response to 'actively becoming mixed up in the solution that produces what water is and what it might be' (p. 235). Calling attention to the ontological practice of bodily engagements with water as 'a kind of participation by which we make different waters rather than making decisions about water' (p. 236), Linton addresses everyday acts of physical engagement with the water fountain as transforming and transformative knowledge practices that challenge the monopoly of bottled water companies. Reflecting on his own experience of drinking from a public water fountain in Canada as a method of practical hydrolectics, Linton draws attention to the different waters made and 'made known through modes of social engagement' (p. 228). Practising hydrolectics through the act of drinking water from the fountain, for Linton, 'can have the potential to change things', including 'physical infrastructures that sustain historical instances of water' (p. 236). The kinds of water and social relations made possible by drinking from a public fountain, rather than drinking branded bottled water from a vending machine, involve the making of 'water as a public good' that '(co)produces public water and a body public' (p. 236).

In recognising direct forms of engagement with the public drinking fountain as a way of knowing water that produces a history of different waters, Linton's Foucauldian genealogical methodology draws attention to mundane methods of everyday bodily engagements with the water fountain as acts of participation. In so doing, Linton's method of practical hydrolectics provides an alternative framework to investigate water and social relations, not through the prism of hydrology or critical intellectual inquiry, but in terms of situated practices that have the potential to change things. Undoing the hegemonic construction of water, in the hope of locating a social understanding of waters and changes in social relations as dialectical and relational processes, involves the repositioning of knowledge of water on two fronts: in the way a person chooses to interact with the public drinking fountain and in the way they make sense of their embodied practices. The empirical project of how to engage the history of water by investigating everyday practices in situ and how technologies for providing drinking water are internally related to particular ontological modes of thought involves ongoing appraisal.

Theorising water

In a review of *What is water?*, the political ecologist Alex Loftus (2011) is sympathetic to Linton's effort to grant nonhumans a role in the historical analysis of water. At the same time, Loftus 'remain[s] agnostic as to whether ANT can really provide the tools for the kind of transformative politics that remains at the heart of Linton's book' (p. 188). While Loftus finds Linton's delineation of the hydrosocial cycle 'path-breaking', outlining the debate in geography between Eric Swyngedouw's (2004) Marxist dialectical approach to water and Linton's ANT Latourian (1993) inspired relational study, Loftus is less optimistic on how engagements with nonhuman others involve the making and transformation of waters. On this point, Loftus quotes Linton from *What is water?* to distinguish between Linton's theoretical relational dialectics and Swyngedouw's dialectical materialist approach for understanding the co-evolution of water and social power. 'Instead of setting on a materialist foundation the process by which history and nature are produced, I argue that it rests on *no particular foundation at all*' (Linton 2010, p. 41, cited in Loftus 2011, p. 187, my emphasis). While Linton is described in this passage as 'critiquing a historical materialist approach' and ultimately ends up accepting 'the common caricature of marxism', Loftus sees Linton as 'contradict[ing] himself', by 'appear[ing] to deny the role of material practices within his theoretical framework' (p. 187). As these quotes suggest, Loftus perceives the difference between Linton's ANT relational approach to waters and a historical materialist perspective, as the latter offering a more profound understanding of how material practices shape the socio-natural world.

On this point, I want to stress there are important differences that require further examination. They can be brought out by tracing Loftus' interpretation of Linton's (2010, p. 41) observation that the history of water 'rests on no particular foundation at all' to support his term the hydrosocial cycle. On the one hand, Linton is indicating that the history of water 'cannot simply be traced as an extended outcome of material relations' (p. 40). On the other, Linton's non-foundational genealogical approach does not involve a commitment to a particular theoretical framework. What Linton is addressing here has little to do with a 'castigation of marxism' as Loftus claims (p. 187), but rather the impossibility of maintaining a rigid separation between material practices and the history of thought. Significantly, in quoting Linton, Loftus neglects to follow Linton's paragraph to the end. After distancing himself from a materialist theoretical framework on the process by which history and nature are produced, Linton continues:

> Instead I take the process itself as foundational and avoid the need to look behind or below it to find a particular driving or determining force. This provides a useful way of analysing both the history of water and how the idea of water articulates with its material and representative forms to produce this history.
>
> (p. 41)

In this paragraph, Linton's concern with analysing the history of water and how it is articulated has little to do with producing a history of thought through ideological frameworks and representational practices. Instead, Linton's problem, similar to Foucault's, is how to engage the history of water without taking the practices that produce thinking for granted. For Linton, ideas and concepts 'cannot be considered to occur independently of their relations with the materiality of non-human nature, the material practices of people, discourse, representation, and so on' (p. 40). It follows, that the problem Loftus sees as 'peculiar' in Linton's non-foundational theoretical framework – as a refusal to conceptualise the relationship between water and society independently of material practices – may indeed 'misinterpret' Linton's method of hydrolectics, as Loftus (p. 187) suggests.

The following discussion examines the theoretical and methodological challenges posed by Linton's method of practical and analytical hydrolectics outlined above. Given the focus of the discipline of geography on the question of what water is, it is perhaps not surprising that we find the most sustained debates on the application of social theory to water in critical human geography. In the article 'The power of water: developing dialogues between Foucault and Gramsci', Ekers and Loftus (2008) show a commitment to demonstrating the benefits of new modes of theorising for gaining an improved understanding into the politics and praxis of everyday life. In what follows, we examine their article more closely. The article presents a detailed argument for an expansion of the role of a Foucauldian approach in conversation with Gramscian theories of power. Ekers and Loftus argue an engagement with both is necessary for developing a theoretical framework for a new research agenda on water. An engagement with these theorists, Foucault and Gramsci, is understood to broaden and specify 'the analytic framework through which the relationship between water and power is interpreted' (p. 708). Drawing attention to Ekers and Loftus' argument, my aim is to further illuminate the challenges of using

Foucault's concept of problematisation for engaging the water fountain as an object to ground theory in everyday hydrosocial practices.

To develop their argument, Ekers and Loftus combine elements of Foucault with Gramsci. In so doing, they seek to address the productive tensions between their works. As the title of the article suggests, what is gained by bringing together the conceptual insights of Gramsci on hegemony and ideology, and Foucault on governmentality and power, is an improved understanding of the 'politics and practicalities of water provision' (p. 699). One of the crucial benefits of connection between Foucault and Gramsci, they argue, is that:

> Both have a deep appreciation for how ideas and types of knowledge are immanent within the materiality of practices and apparatus. Thus, Gramsci and Foucault can provide us with an appreciation of how specific rationalities of government and ideologies are internalised within … cities, and for our purposes, produced waterscapes. Moreover, from a Foucauldian perspective, this entails managing the conduct of people and their relations with the material world, customs, beliefs, and ways of acting and thinking. From a Gramscian perspective, water infrastructures can be considered part of the hegemonic apparatus through which forms of 'common sense', in support of a specific group's interest, come to be constituted.
>
> (p. 706)

Despite their insistence that the theoretical differences and connections between Gramsci and Foucault represent productive tensions, Ekers and Loftus agree with the late urban geographer Clive Barnett (2005) that these tensions 'must be made explicit' and 'not finessed away' (p. 713). Appraising the capacity of a Gramscian or Foucauldian approach for dealing with the politics of water, they argue, 'must surely be more than a theoretical exercise. Rather it must depend on the ability of the framework to account for the untidiness and political struggles of everyday life' (p. 714).

In addition to emphasising the analytical resources made available by converging parts of Gramsci and Foucault to address political struggles of everyday life, Ekers and Loftus make some methodological suggestion on how these differences and tensions 'may provide conceptual tools through which the water research agenda can be enlivened' (p. 701). The goal of exploring these tensions is to reinvigorate theoretical debates and 'develop a research agenda around the quotidian ways in which water and politics are intertwined' (p. 713). By bringing Gramsci's theorisations of everyday practices together with Foucault's concern with knowledge and power, Ekers and Loftus call for more attention to specific practices and 'the micropolitics of things like urban water provision through which people are subjected in specific ways' (p. 713). This includes the changing practices of everyday access to water and everyday interactions, and the 'practices and techniques through which social transformations are enacted' (p. 699). For Ekers and Loftus, the greatest obstacle to pursuing this type of research 'is navigating the tricky methodological terrain'. But paying attention to the broad political, economic, cultural, and economic relations, and understanding 'how they are articulated, expressed and resisted in the ideologies and techniques of power of everyday life', they stress, 'is no easy task' (p. 712).

In summing up what their effort of combining elements of Foucault with Gramsci add to studies of water, Ekers and Loftus suggest their heuristic approach brings to the table a set of conceptual and political resources 'through which we can gather up the provocative threads of research already undertaken and take them forward through a nuanced analytical framework' (p. 713). The Gramsci–Foucault problematic is seen to hold special promise for research already undertaken, as it serves 'to explicate many of the crucial insights that have largely remained implicit in the water literature' (p. 713), and illustrate 'threads of research closely akin to the concerns that preoccupied Gramsci and Foucault' (p. 712). In what follows, we consider how Ekers and Loftus' rethinking of social theory for water-related studies helps us understand crucial insights that remain implicit in Linton's hydrosocial research on water, and concerns that preoccupied Foucault.

Problematising theory: public water fountains as world-making spaces

As we have seen in this chapter, Linton's (2010) question *What is water?* has given rise to new calls on the uses of Foucault's social theory in the discipline of human geography, and to claims that Linton's relational ANT-inspired Foucauldian approach does not quite fill the gap on how to engage material practices in the history of thought about water. Returning to and reclaiming Linton's concept of the hydrosocial cycle and method of hydrolectics, I want to suggest offers a useful approach not only for understanding ways of thinking with water but developing new ways of engaging with the water fountain in the conduct of research. Linton's methods of analytical and practical hydrolectics offer a particularly useful example for rethinking Foucault's (1984) concept of problematisation in relation to the hydrosocial cycle. In tracing the history of modern water as the effect of historical processes, social practices of drinking, and political struggles over access to public water, Linton's method of hydrolectics engages problematisation as *both* a mode of critical thought and object of analysis. On the one hand, using figures of the water fountain and vending machine to illustrate how people relate to water, Linton's method of analytical hydrolectics involves the task of narrating and tracing the hydrosocial cycle and reflectively thinking about the historical and social formation of waters. Linton's photographic illustrations of the water fountain being displaced by the vending machine in the Canadian context identifies a range of actors that make, interrupt, and transform the hydrosocial cycle. The alternative way of engaging with water through bodily practice, made available by Linton's method of practical hydrolectics, opens the investigation on what is water to an embodied situated common sense making practice that does not leave the task of inquiry about how waters are made and known to social theorists and social scientists. Practising hydrolectics insists that the process of thinking with water in everyday urban contexts can only be undertaken through a process of engagement. Here then, there is an implicit acknowledgement of problematisation as a situated practice of common sense making to which Linton attaches meaning through the process of participation. While the aim of investigating the hydrosocial cycle is to get a better understanding on how water becomes the object, site, and substance of thinking, on the question of *how* different waters are made, there are no empirical sources other than Linton's own personal experience. Linton's own personal first-person problem description of using the public water fountain in Canada (as opposed to a vending machine) is provided to illustrate the relationship between a participant observer and problematic situation.

One of the purposes of Gay Hawkins' work on plastic bottled water is to address this empirical gap. The chapter 'Mobile drinking: mobile water practices and ontological politics' from 2019 contains an attempt to move beyond a critique of plastic bottled water and open up an empirical approach to understanding everyday drinking practices. Investigating the rise and effects of carrying and sipping from plastic water bottles, Hawkins invites a reassessment of Linton's method of hydrolectics. In her evaluation of mobile drinking as a social relation, Hawkins addresses the impact of the growth of vending machines and the declining use of the water fountain. In contrast to Linton's performance of the hydrosocial cycle, which distinguishes sharply between the use of public water fountains and vending machines, Hawkins analysis of mobile drinking suggests a disadvantage of Linton's method of analytical hydrolectics, is that it falls into the trap he wishes to avoid. In Linton's history of water, access to free drinking water in cities is attributed to the public drinking fountain, and the disruption of this cycle to the diversion of a 'social-natural process' through commercial bottled water vending machines (Linton and Budds 2014, p. 175). For Hawkins, this deficit model of analysis reduces changes in everyday drinking practices to an 'opportunistic logic of the corporation and decline in water infrastructures and state provision' (p. 118). Understanding the relations between drinking practices enacted by vending machines and those associated with taps 'cannot be reduced to a crude opposition between markets and states or private and public; many mobile drinkers do not understand or justify this practice in these terms', Hawkins argues (p. 111). In Hawkins research, Linton's methodological reassessment of water and the social is seen to be without substance. Drawing on empirical findings from her own study of bottled water practices in Australia, with Emily Potter and Kane Race (Hawkins, Potter and Race 2015; Race 2012), Hawkins shows how concerns about health, water quality, and hydration, had far more influence over the shift to new mobile drinking practices than the commodification of water and disappearance of public water fountains. The idea of a causal relation between the disappearing access to public drinking water and the rise of a market in single-serve branded bottles, is thus, Hawkins argues 'difficult to sustain' (p. 118). But in investigating the practice of reaching for the bottle as provoking new modes of drinking and new water realities, there is a danger of reifying the water fountain as a devalued thing that doesn't need explaining. While Hawkins makes clear the changing infrastructure of public drinking water fountains in urban contexts is beyond the scope of her analysis of mobile drinking, in what follows we consider how an engagement with the water fountain and public drinking practices might be developed through Foucault's concept of problematisation.

Problematisation and urban inquiry

The following discussion enables us to examine the impact of Foucault's notion of problematisation for understanding practices of drinking in Linton's hydrosocial cycle more closely. As we have seen in this chapter, the effort afforded to revelling in the tensions and resonances between a neo-Marxist and Foucauldian approach by Ekers and Loftus, and an STS-inflected approach by Hawkins and Linton, is perceived to develop new water research agendas. In this next section, we turn to the approach of Barnett (2005, p. 9), who warned against 'the blind spots' in both Marxist and Foucauldian governmental theories of power for analysing 'how top-down initiatives "take" in everyday situation and routines'. Similar to Linton, and Ekers and Loftus, Barnett (2015) and Barnett and Gary Bridge (2017) refer to Foucault as a theorist for developing urban inquiry. Here, I focus on their analysis of concepts of urban inquiry, which presents a detailed argument for an engagement with Foucault's work and the challenges it poses for addressing everyday urban waterscapes, and what this means for social theory and social research.

A central preoccupation of Barnett and Bridge is to challenge the assumption that Foucault's main contribution to social theory 'lies in conceptualisations of power' (p. 1190). Calling for a new approach to Foucault's social theory in critical urban inquiry, Barnett and Bridge seek to resist the demand placed on human geographers to produce more coherent conceptualisations of issues, through either the clarification of concepts and a commitment to correcting the usage of concepts or 'ideological subterfuge requiring exposure' (p. 1188). Whilst Foucault's ideas have often been cited as the authoritative voice for a theory of power, Barnett and Bridge point out critical engagements presume concepts for urban inquiry 'remain the special preserve of a cadre of intellectuals' (p. 1201). Although they agree that key concepts of human geography and related fields are thoroughly problematic, Barnett and Bridge propose an alternative approach to using Foucault's theory. Concepts of urban inquiry are developed not through a process of academic correction but reference to Foucault's idea of problematisation. In taking Foucault's work on problematisation as a starting point to expand debates about the status of concepts in urban inquiry, Barnett and Bridge call into question the adequacy of existing understandings of problematisation as the task of 'critical de-familiarisation' (p. 1201). The problem with this academic activity, they argue, is that it works by putting things beyond question and doesn't appreciate 'everyday modes of problematisation' (p. 1194). Thinking problematically about the concepts of urban inquiry, as an engaged action across any number of fields of practice, Barnett and Bridge interrupt conventional academic interpretations of what Foucault's concept of problematisation offers social theory. Developing a methodological perspective on the ordinariness of problematisation which takes seriously Foucault's genealogical method of inquiry, Barnett and Bridge seek to open up urban studies to a 'broader range of sites and sources for theoretical reflection and insight' (p. 1201).

In opening an alternate way of engaging with everyday modes of problematisation as a dispersed dimension of practices, Barnett and Bridge offer a useful extension of Foucault's genealogical methodology for broadening, challenging, and transforming modes of inquiry on urban drinking practices. Their intervention is helpful for providing the discipline of urban geography and social sciences more broadly with alternative ways of engaging with Linton's hydrosocial paradigm and his example of the mundane technology of the water fountain. The challenge of thinking problematically with Linton's concept of the hydrosocial cycle is not solely a refined academic analytical skill or the preserve of social theorists. Taking seriously Barnett and Bridge's methodological perspective on the ordinariness of problematisation, in what follows we consider how Linton's contrasting methods of analytical and practical hydrolectics might be further delineated by reference to Foucault's work.

For Barnett and Bridge, and Barnett (2015), the starting point for engaging the activity of problematisation in urban inquiry entails resisting the temptation to use it as either a critical method of analysis or procedure to be followed by researchers. On the one hand, Barnett and Bridge question the interpretation of problematisation in critical policy analysis, as a means for understanding processes of problem formation (e.g. Bacchi 2012). Referencing Carol Bacchi's (2009) Foucauldian-influenced methodology, of investigating 'what's the problem meant to be', Barnett and Bridge seek to resist this constructionist tradition of questioning taken-for-granted understandings of policy problems and revealing the operations of power and processes of subject formation. On the other, Barnett and Bridge question the procedure of problematisation followed by theorists using ANT and governmentality studies. Here, problematisation involves tracing 'the process of calculated, strategic interventions through which problems are defined in ways that enrol various partners and shape subsequent pathways of action, decision and inquiry' (p. 1191). Arguing against these two distinct types of analyses of problematisation, Barnett and Bridge suggest the activity of problematisation as a practice of reflection is not a 'refined academic skill' but a 'much more widely distributed feature of urban living' (p. 1201).

In what follows we consider the implications of Barnett's, and Barnett and Bridge's understanding of problematisation not as a critical method or object of analysis but following Colin Koopman (2013) and Paul Rabinow's (2011, 2012) synthesis of Foucault's ideas as an object of inquiry for rethinking the activity of problematisation in Linton's practice of hydrolectics. For Barnett, the relevance of problematisation for social inquiry involves asking; '"how and why certain things (behaviour, phenomena, processes) become a *problem*"' (Foucault 2001, p.171 cited in Barnett 2015, emphasis in original). To distinguish Foucault's interest into the history of thought, Barnett quotes Foucault (1984) from an interview with Paul Rabinow on 'Polemics, Politics and Problematisations' at length:

> 'Thought is not what inhabits a certain conduct and gives it its meaning: rather, it is what allows one to step back from this way of acting or reacting, to present it to oneself as an object of thought and question it as to its meaning, its conditions and goals' (p. 388).

Following Foucault, Barnett points out the application of problematisation for social inquiry should not think of problematisations as being 'mere representations of pre-existent conditions' (p. 58). Applied to Linton's method of hydrolectics, the task of inquiry is to analyse how and why drinking from the water fountain becomes a problem of thought within an unproblematic field. Addressing such questions, for Barnett, depends on a descriptive mode of analysis that 'helps draw into view the significance of the difficulties and concerns that already animate people's actions' (p. 58). This descriptive approach requires a method of what Paul Rabinow (2012) calls problem description; diagnosing formerly stable figures, scientific knowledge, and assemblages, and how they are being recombined and reconfigured in a problematic state of affairs 'as simultaneously the object, the site, and ultimately the substance, of thinking' (p. 32). The primary concern of the analyst for practising an inquiry on contemporary problems and possible solutions, and how they become available to thought, involves inventing means for observing the 'relationship between observer and problematised situation' (Rabinow 2002, p. 139).

In the concluding section of this chapter, we consider how Martha Kaplan's (2011) ethnography of the drinking fountain and water cooler goes some way to responding to the challenge of inventing methods to observe, analyse, and become directly engaged in the relationship between an observer and problematic situation. In so doing, we address how Kaplan's empirical study of the water fountain and water cooler complicate Linton's understanding of the decline and disappearance of free public drinking fountains as a direct consequence of vending machines. As we have seen in this chapter, Linton's method of analytical hydrolectics involves tracing and narrating the relation between two sets of competing physical infrastructures, bottled water from a vending machine and water from a fountain. Problematisation involves an encounter between these two sets of infrastructures. This contrasts with the construction of waters in Linton's method of practical hydrolectics, which brings into focus the description of a participant observer doing something with the water fountain that provokes thinking in ways that reconfigure the hydrosocial cycle. In this way, Linton's practice of hydrolectics offers a more empirically robust methodology for investigating hydrosocial relations in everyday contexts that expand engagements with the ordinariness of problematisation as separate from theory. However, whilst Linton's historical analysis of thought in relation to water involves an engagement with the water fountain and everyday drinking practices, it does not involve an investigation of public water fountains and their users. In the article 'Lonely drinking fountains and comforting coolers: paradoxes of water value and ironies on water use', Kaplan (2011) elevates the neglected water fountain from a minor actor to a major protagonist in the study of technologies of drinking practices and the situated meanings of water use. Taking the mundane technology of the water fountain as the object of inquiry, Kaplan switches her theoretical lens from a critical Marxist analysis of the decline of the water fountain as reflecting water commoditisation to ANT and the work of Latour (1996, 1999) to consider the relations between publics and things. At the outset of Kaplan's research, she acknowledges the neglected and rejected water fountain. Kaplan's investigation seeks to address how the fall in its use and changing value are made within everyday practice. Using the methods of observation, interviewing, and photography, Kaplan undertakes ethnographic research between 2006 and 2008 with water sources and water drinkers in ten college campus buildings in New York.

In her study, Kaplan traces differences in the lives of drinking fountains, single-serve bottles, water coolers, and their social relations. The results were surprising, especially in terms of the words selected by participants to describe drinking fountains compared with water coolers. While people spoke of drinking fountains as sources of dangerous, polluted, medicated water with chlorine or fluoride, they tended to avoid common words and ordinary nouns to describe the water fountain, concentrating instead on a vocabulary of smell and taste. Kaplan observers the word 'water fountain' was not only disowned with people 'refusing to know its name' (p. 520), but 'few called the water supplied by the drinking fountains "public"' (p. 521). Kaplan's STS-inspired ethnographic investigation of drinking fountains involved a comparison of the water dispensed through the fountain and vending machine, with the object of the water cooler. Asking questions of the relations between technology and everyday drinking practices ethnographically, Kaplan found that the water fountain was not supplanted by single-use plastic water bottles, but spring water coolers 'hybrids of public and private, individualisable but free-flowing hydration' (p. 534). By substituting critical theoretical analyses of water technologies with an ethnographic method, words could be captured to describe the everyday realities of the water cooler. In contrast to the water fountain, water coolers were described as things that assemble 'collective care', 'community', and 'trust' (p. 534). In offering an empirical methodology for engaging the question of what water is in everyday practices, Kaplan's STS case study of the water cooler and drinking fountain provides a missing dimension to Linton's method of practical hydrolectics. In Kaplan's research, the implications of a more sustained empirical application of an ANT approach to the study of water is highlighted. But in combining Latourian concepts with an ethnographic empirical agenda, the question of how the water cooler and drinking fountain become objects of thought *in practice* becomes sidelined. Investigating the problematisation of plastic bottled water in everyday practice, and how the water fountain provokes thinking in specific settings, requires more attention to this mundane technology and common sense making practices as empirical questions that cannot be explained away, issues we address in greater detail in Chapters 3, 4 and 5. Before developing a proposal for doing ethnographic research with London's #OneLess refill water fountain pilot, in the next chapter we look at how the social sciences more broadly have responded to the environmental issues of the plastics and climate crises and grappled with the methodological question of how to engage with the impact of environmental problems upon social processes, in ways that trouble any straightforward problematisation of everyday drinking practices.

References

Bacchi, C. (2009) *Analysing policy: What's the problem represented to be?* Pearson.

Bacchi, C. (2012) Why study problematisations? Making politics visible. *Open Journal of Political Science* 2(1), 1–8.

Barnett, C. (2005) The consolidations of 'neoliberalism'. *Geoforum* 36(1), 7–12.

Barnett, C. (2015) On problematization: Elaborations on a theme in 'Late Foucault'. *Nonsite*. Issue 16, June 22.

Barnett, C. and Bridge, G. (2017) The situation of urban inquiry: Thinking problematically about the city. *International Journal of Urban and Regional Research* 40(6), 1186–1204.

Budds, J., Linton, J. and McDonnell, R. (2014) The hydrosocial cycle. *Geoforum* 57, 167–169.

Eckers, M. and Loftus, A. (2008) The power of water: Developing dialogues between Foucault and Gramsci. *Environment and Planning D: Society and Space* 26(4), 698–718.

Foucault, M. (1984) Polemics, politics and problematizations. In P. Rabinow [Ed.] *The Foucault reader* (pp. 381–390). Pantheon Books.

Foucault, M. (2001) *Fearless speech*. Edited by J. Pearson. Semiotext(e).

Hawkins, G. (2019) Mobile drinking: Mobile water practices and ontological politics. In C. Maller and Y. Strengers [Eds.] *Social practices and dynamic non-humans: Nature, materials and technologies* (pp. 109–128). Palgrave.

Hawkins, G., Potter, E. and Race, K. (2015) *Plastic water: The social and material life of bottled water*. MIT Press.

Kaplan, M. (2011) Lonely drinking fountains and comforting coolers: Paradoxes of water value and ironies on water use. *Cultural Anthropology* 26(4), 514–541.

Koopman, C. (2013) *Genealogy as critique: Foucault and the problems of modernity*. Indiana University Press.

Latour, B. (1993) *We have never been modern*. Harvard University Press.

Latour, B. (1996) *Aramis, or the love of technology*. Harvard University Press.

Latour, B. (1999) On recalling ANT. In J. Law and J. Hassard [Eds.] *Actor network theory and after* (pp. 15–25). Blackwell.

Linton, J. (2010) *What is water? The history of a modern abstraction.* University of British Columbia Press.

Linton, J. and Budds, J. (2014) The hydrosocial cycle: Defining and mobilising a relational-dialectical approach to water. *Geoforum* 57, 170–180.

Loftus, A. (2011) Thinking relationally about water: Review based on Linton's 'What is water?'. *The Geographical Journal* 177(2), 186–188.

Rabinow, P. (2002) Midst anthropology's problems. *Cultural Anthropology* 17(2), 135–149.

Rabinow, P. (2011) Dewey and Foucault: What's the problem? *Foucault Studies* 11, 11–19.

Rabinow, P. (2012) How to submit to inquiry: Dewey and Foucault. *The Pluralist* 7(3), 25–37.

Race, K. (2012) Frequent sipping: Bottled water, the will to health and the subject of hydration. *Body & Society* 18(3-4), 72–98.

Swyngedouw, E. (2004) *Social power and the urbanisation of water: Flows of power.* Oxford University Press.

2 Probing problematisation in the plastics and climate crises

Before describing my methodology for doing social inquiry on the refill water fountain – through the study of #OneLess' campaign in Chapter 3, and more programmatically in relation to the #OneLess refill water fountain pilot in Chapters 4 and 5 – in this chapter we will look more closely at Gay Hawkins, Emily Potter and Kane Race's (2015) investigation of plastic water as a case study to open up insights into the conceptual, methodological, and practical significance of what problematisation offers for social research on campaigns to reinvent the water fountain. To this end, the chapter provokes both reflection on how social theorists and social researchers have themselves grappled with the implications of problematisation for researching drinking practices and campaigns to contest plastic water, and their capacity to imagine a new sense of purpose. To do so, the chapter takes Hawkins et al.'s research on mobile drinking practices as an entry point to investigate theoretical and methodological approaches that enable and prevent thinking with water more problematically. By putting into dialogue the different conceptual stances and methodological perspectives deployed by academic scholars, including philosophers, sociologists and STS-influenced social scientists, the chapter provides an overview of contrasting contributions to thinking with environmental problems associated with the plastics and climate crises, and the strengths and limitations of different approaches to engage global environmental issues, including; social practice (Shove 2010a,b,c, 2012; Shove, Pantzar and Watson 2012), material participation (Marres 2012, 2015), apocalyptic imaginaries (Swyngedouw 2010), metamorphosis (Beck 2015, 2016), an ecology of practices (Stengers 2005), and an ecology of futures (Michael 2017). In doing so, the

DOI: 10.4324/9781003616757-3

chapter addresses how questions and solutions to the plastics and climate crises have been imagined and how they might be possible to re-imagine in methodological and practical terms of problematisation.

Problematising plastic water

In a number of linked articles and the co-authored monograph *Plastic water: The social and material life of bottled water*, Hawkins et al. (2015) and Hawkins (2010, 2011, 2013, 2018, 2019) have called for a new empirical approach to disposable plastic bottled water that gets beyond activist critiques. What concerns Hawkins, is the tendency for activist campaigns to 'ridicule the sipper or demonise the bottle', without paying attention to mobile drinking and how bottled water practices are practically assembled. Drinkers, Hawkins points out, 'use bottles in various unexpected ways' (2019, pp. 110, 112). The danger with framing plastic water in terms of fixed attributes and negative qualities, is 'to exclude [the plastic bottle] from the active realm of the social' (2018, p. 340). By switching the focus of inquiry from the demonised plastic bottle to bottled water practices, Hawkins (2019, p.111) addresses how plastic bottles 'are accounted for and enacted in practices – and with what effects?'. Investigating the multiple ontological and political realities of carrying and sipping bottled plastic water on the move, Hawkins' (2019, p. 112) approach describes the practice of mobile drinking as involving 'a complex association of elements, materials, knowledges and performances – human and non-human, economic and non-economic'. In so doing, Hawkins outlines the difference between her ANT and STS informed approach with that of Elizabeth Shove, Mika Pantzar, and Matt Watson's (2012) practice theory.

The purpose of Shove et al.'s (2012) practice theoretical research is to provide a conceptual and methodological framework to evaluate policy interventions designed to address the engendering of more sustainable routines and habits. What this approach involves in practice is developing 'other sorts of data and other styles of enquiry'. These might include quantifying the 'growth of certain practices and the demise or transformation of others' (Shove et al. 2012, p. 115). In highlighting the implications of practice theory for social scientific research methodologies on climate change policy, Shove advocates changing the terms in which problems are cast from a language of 'outcome of individual choice' to 'what people do' as part of a process 'of sociotechnical change' (2010c, p. 1.12). The advantage of placing social theories of practice 'centre stage' in social enquiry on sustainability and environmental issues is to imagine different aspects of what 'sustainable practice might entail' and how such change might come about (Shove and Spurling 2013, pp. 3, 2). In order to conceptualise new habits of thought, Shove (2012) shifts her thinking about habits from the theorist's point of view to the 'habits point of view'. By treating habit as a mode of enactment that is performatively made and broken in practice, Shove develops a practice-centric method for engaging habits in everyday contexts. In moving the focus of inquiry from individual habits to 'habit-demanding practices', Shove recognises materials, meaning, and competence must 'be readily and reliably available' for habit changing agendas to materialise (pp. 101, 110, 110).

Reviewing the role of materials and resources in configuring practice in Shove et al.'s (2012) work, Hawkins (2019) questions Shove et al.'s criticisms of ANT for 'diverting attention from more ordinary questions about what these cyborg/hybrid entities are actually *doing*' (Shove et al. 2012, p. 10, cited in Hawkins 2019, p. 113 emphasis in original). For Hawkins, Shove et al.'s critique of ANT and STS, of not attending to the question of what objects do *in* practice, misrepresents analyses of the nonhuman and the dynamics of assembling the social. Using her case study of the material and technical capacity of the plastic bottled water to practically enact 'new codes of self-care' and establish a new practice and drinking reality, Hawkins (2019, p. 119) responds to Shove et al.'s critical assessment of ANT and STS. By acknowledging how plastic bottled water practices are implicated in the making of ontologies, Hawkins' rejoinder to Shove et al. holds considerable interest for social research on the effects of environmental campaigns to contest plastic bottled water. Hawkins' response provides a means to revisit the empirical problem of how to engage everyday public drinking practices from an STS methodological perspective. Reflecting on Hawkins' contribution to this debate, in this chapter we revisit the 'methodological dilemmas' (Hawkins 2018, p. 340) of engaging controversies on plastic bottled water within social inquiry and take seriously Foucault's (1984) notion of problematisation as a means to further open the debate Hawkins sought to undertake.

In Hawkins et al.'s (2015) detailed historical study of the social and material life of plastic water, they trace the methods and techniques that frame mobile drinking practice as both a 'problematized object' of activist critique, and 'something that can acquire lively "powers of engagement"'. Investigating the practices of anti-bottled water activism campaigns to transform single-use bottled water into a controversial thing and trigger 'new relations, problematizations, and identities for this once innocent commodity', Hawkins et al. empirically track 'how bottled water was made into an issue' (pp. xxi, xxii, 180, 147). Focusing their investigation on the methods and devices of activism, to '*provoke* a political reality for bottled water' (Hawkins 2018, p. 341, emphasis in original), Hawkins et al. (2015) address '*how* activism is enacted', how these campaigns work practically to situate controversy, and how activists transformed plastic bottles into 'not just a political problem but a political material', with the capacity to provoke and engage drinkers. Examining the antagonistic methods and devices of various political campaigns to singularise the mundane disposable plastic water bottle into a 'matter of concern', Hawkins et al.'s research draws attention to the troubling effects of specific requalification practices to contest and 'problematize bottled water' in relation to plastic waste and pollution, and environmental concerns (pp. 151, emphasis in original, 151, xvii, 159).

Documenting the qualifying and requalifying practices of bottled water controversies as a central object of inquiry, Hawkins (2018) highlight how situating objects, methods, and settings in the research process cannot be clearly separated. The problem with much activist analyses of the bottled water industry, argues Hawkins, is they represent the bottle as a 'passive object whose political and environmental impacts are written in advance' (2011, p. 544). What's missing from these activist accounts is a sense of how the plastic bottle – as an object – participates in contexts of its use in dynamic ways. Focusing on the publics called into being by the recycling logo on plastic bottled water, Hawkins (2011) draws attention to plastic water as an object critical to the materialisation of publics. Looking beyond activist analyses of the problematisation of packaging is necessary, argues Hawkins, to see how plastic bottles are constitutive of '*everyday* or ordinary forms of political engagement and participation' (2013, p. 78, emphasis in original). These everyday material dynamics of participation reveal the ways in which plastic water bottles can 'open up a virtual space for a potential or "infrapublic" to engage in political practices'. This 'might be as minor as refilling the bottle', 'or as major as starting a national campaign to re-establish public water fountains' (Hawkins 2011, p. 551).

Hawkins et al.'s (2015) empirical analysis of the material dynamics of participation draws into focus the role and methods of non-government organisations to make the problem of bottled water public. Comparing the framing of plastic water in anti-bottled water media campaigns, to the 2009 'Do Something' water fountain campaign, an experiment initiated by Manly council to reintroduce and reinvent public drinking water fountains in Sydney, Australia, Hawkins et al. describe the 'Do Something' campaign as a *'disruptive event'* that sought to 'displace the bottle by *revaluing* public water' (pp. 180, 178, emphases in original). In so doing, Hawkins et al. draw attention to this intervention to install new public water fountains, which included in some instances a refill station to facilitate refilling of pedestrians' own water bottles, as the most effective requalifying strategy to challenge the practice of buying and carrying bottled water. By appealing to the ordinary act of drinking water, the 'Do Something' campaign is understood to have 'promoted a re-evaluation of the water fountain as a "public thing", a democratic object shared by all' that 'offered a real alternative to single-use purchasing of plastic water (pp. 150, 176). Drawing on Noortge Marres' (2012) STS empirical analysis of political participation as involving things, devices, and settings that engage publics in political issues, Hawkins et al. show how bottled water became a 'political object' and the specific devices that are drawn on to 'contest its normalization' in daily life (pp. 150, 214).

In her empirical study of material participation in environmental issues, Marres (2012, 2015) shifts the focus of inquiry from theoretical interpretations of environmental problems by academics to empirical descriptions of the deployment of everyday objects and ordinary situations to produce disruptions of social life. Thinking with everyday interactions with sustainable living devices as a 'particular type of *sociological* experiment', Marres shows how the complex problems of climate change are made available for 'practical intervention by everyday people'. Describing the disruption of established ways of doing things as an '"object-centred"' sociological method, Marres likens sustainable living experiments to the breaching experiments developed by the enthnomethodologist Harold Garfinkel in the 1960s to generate insights into everyday life (Garfinkel 1967/1984), methods we return to in Chapter 5. For Marres, social scientists' contribution to contemporary debates on sustainable living involves a positive affirmation of the empirical devices that facilitate participation in environmental issues. The crucial question concerns the participatory capacities of problematising objects and devices in the distribution of problems. The methodological dilemma concerns how to make the everyday objects of material participation in sustainable living experiments describable in situ. The task for social research is 'to grasp the effects of interventions especially those that tend to go unacknowledged and remain invisible' (p. 93, emphasis in original, xii, 94, 102).

What distinguishes Marres' methodological perspective from Shove et al.'s practice-centric approach is the task of grasping the politics of objects and the technologies and settings 'which enable them to operate'. Situating the domestic home 'as a *critical* site of research', Marres draws into focus the controlled disruption of ordinary situations as transformative of everyday objects into a focus of public attention, and familiar contexts into engaging environments. In drawing attention to the mundane household technologies and domestic settings which enable problematisation as a 'practical accomplishment', Marres shows how sustainable living experiments involve the 'modification of habits and habitats'. In order to rethink the role of social research in everyday living experiments and the particular role of social researchers, Marres suggests researchers 'investigate how particular devices make possible the investment of things with political capacities'. Avoiding an affirmative approach to material centred public experiments, Marres calls for social scientists to develop 'a more nuanced and ambivalent appreciation of the role of things in the enactment of public participation' (2015, pp. 107, 25, 152, 86, 107, 9). By drawing into focus everyday engagements with global environmental issues in local domestic contexts as a practical accomplishment, Marres shows why studying problematisation in local settings is an important 'site for the performance of "socio-ontological" change' (2015, p. 89) that gives social scientists an entry point to investigate environmental participation empirically. Studying the use of experimental methods as a particular form of knowledge production is seen to enable social scientists to 'explore collective practices of researching social, cultural, and technological change, as engaged in by actors who do not necessarily identify themselves as "social researchers"'. However, Marres (2012) warns it is a mistake to assume the sociologist is the 'sole agent of social research in this situation'. The collection and analysis of social data she argues is 'performed by a whole range of actors and devices'. What distinguishes the contribution of social science perspectives is the task of 'adding its techniques and modes of research and analysis to those already deployed' so that 'social researchers may render or foreground different aspects of it' (pp. 111, 124, 111–112, 124).

In documenting the reinvention of the public drinking fountain as a strategy to reduce plastic bottled water consumption and address issues of climate change, Hawkins et al.'s description of the 'Do Something' campaign opens up the question of the role, capacity, and contribution of social researchers to respond to the challenge of researching experiments is sustainable living (Marres 2012; see also Marres, Guggenheim and Wilkie 2018 and Kimbell, Guggenheim and Marres 2018). Whilst the main focus of Hawkins et al.'s study involves the analysis of *how* anti-bottled critique and activism practically made plastic water a political problem, and the representational methods and devices enacted by activists to make a political reality for bottled water, in their description of the 'Do Something' campaign, Hawkins et al. foreground a way of attending to the mundane technology of the refill water fountain. The 'Do Something' water fountain campaign is described as a positive experiment, 'something that posed questions' to the public and 'invited reflection on that taken-for-granted experience'. In drawing attention to the 'Do Something' experiment as a positive strategy for engaging the social dynamics of environmental issues of climate change and plastic bottles in everyday practice, Hawkins et al.'s investigation of the qualifying and requalifying practices of the anti-bottle water activists to enact controversy and provoke a new social reality foregrounds not just the disruptive techniques and devices used to contest bottled water, but *ordinary modes of problematisation* as involving embodied engagements with the new water fountain that made being and thinking in common a new reality (pp. 181, 180).

This move contrasts with Erik Swyngedouw's (2010) more pessimistic approach to environmental knowledge practices. Responding to the elevation of knowledge of the global problem of climate change onto the terrain of public concern and policy, Swyngedouw engages in an epistemological critique of the staging of environmental problems and their political effects. Calling into question the framing of the climate conundrum through the invocation of fear and danger, Swyngedouw takes as his object of analysis the negative rhetoric in representational and discursive strategies which frame the environmentally apocalyptic future. By positioning the public as universal helpless victim of an ambiguous unmanned threat beyond the state and our control, Swyngedouw argues fearful apocalyptic imaginaries of climate change de-politicise power relations within the social system. The political construction of universal sufferers is interpreted as a form of populist governing that invokes a socially homogenised common humanity and blames problems on an outsider. Within this post-political public sphere, the empty signifiers of climate change policy replace proper politics. Doing politics is reduced to policymaking by non-state and quasi-state institutional experts and stakeholders, who 'operate with a generally accepted consensus of a global and largely (neo)liberal capitalism, the right of individual choice, an ecological awareness and the necessity to continue this, to sustain the state of the situation' (p. 227). Foregrounding the consequences of fear tactics of global environmental governance as de-politicising publics, problems, and democratic politics, Swyngedouw's political economy critique draws on Alain Badiou's (2008) analysis of 'ecology as the new opium for the masses' (p. 219).

Swyngedouw's critical political analysis of the climate conundrum contrasts with the more optimistic conceptualisation of the climate crisis by the late Ulrick Beck, as not climate change, but 'much more and something very different' (2015, p. 79). In Beck's analysis of climate change as neither apocalypse nor progress but 'something "in-between"', he proposes a methodological approach developed through his theory of metamorphosis. Metamorphosis, argues Beck, 'is happening through everyday practices on all levels; and it is effecting everyone everywhere differently' (pp. 75, 77). By switching the focus of his analysis from the negative apocalyptic side effects of climate change to the positive emancipatory side effects of global risk, Beck calls for social theorists and sociologists to recognise *how* we are becoming different. What's required to engage the metamorphosis of the world in thought, argues Beck, is a methodological perspective that is open to thinking, imagining, and knowing how people get involved in different situations, learn, experience, and get through change in their everyday life in unexpected ways on a range of levels. What gets erased from social theory and political critiques of climate change and society as a threatening future, argues Beck, is understanding 'the change of the conditions and understandings of change' (2015, p. 76). In an interview with Gabe Mythen, Beck stresses a major methodological problem for social researchers is to 'create terms which minimize the risk of misunderstanding' the mode of change (Mythen 2018a, pp. 181). Whilst Beck's theory of metamorphosis is welcomed as an invitation to stimulate the imagination to produce original concepts and experiment with new tools of analysis capable of describing the complexities of social change and a common sense of problems, Mythen argues social scientists are left with a 'methodological conundrum'. The empirical challenge of engaging the everyday quality of metamorphosis habitually and routinely in everyday practices requires the individual undergo a 'form of personal metamorphosis in order to view the world with fresh eyes'. Theorising transformations in everyday processes through the lens of metamorphosis is considered a 'formidable' task (Mythen 2018b pp. 198, 199, 199), which for Beck involves 'disrupting the settled status quo of knowledge production' (Mythen 2020, p. 389).

Commenting on the challenges of Beck and Swyngedouw's contrasting methodological and epistemological approaches for understanding, engaging, and intervening in debates on society, the climate crisis, and the environment, the late John Urry (2011) underscores the capacity of emancipatory catastrophism and apocalyptic post-political populism for social theory and sociology. Overall, Urry perceives Beck's cosmopolitan vision of the staging of fearful risks on a global scale as expanding the empirical into the world in unexpected ways, somewhat convincing. At the same time, Urry finds Beck's theory of metamorphosis 'insufficiently precise' about the character of social practices. What troubles Urry about Swyngedouw's critical political thesis of ecology as the new consensual '"opium for the masses"' is a nostalgic golden age view of proper politics, and little reference to the social science of climate change (pp. 11, 91). Whilst Urry's criticisms of Swyngedouw's critical epistemology of climate change and Beck's theory of metamorphosis highlight the strengths and limitations of these contrasting theoretical and methodological approaches to the climate crisis, Isabelle Stengers' (2015) evaluation of academic critique and everyday practice suggests the metaphor of the pharmakon and concept of metamorphosis may be more relevant for the level of methodological transformation required for understanding how to respond to the climate and plastics crises than Urry's criticisms of apocalyptic and emancipatory futures suggest.

Reclaiming problematisation

In catastrophic times: Resisting the coming barbarism, Stengers (2015) claims one of the biggest academic challenges presented by the climate change question to academics concerns the matter of thinking and an ability to think differently. Following Alfred Whitehead's (1968) plea for an association of the future, 'neither with the advance of knowledge nor with progress but rather with radical uncertainty', and Donna Haraway's (1988) commitment to producing situated knowledge that is 'indebted to the existence of others who ask different questions, importing them into the situation differently, relating to the situation in a way that resists appropriation in the name of any kind of abstract ideal' (Stengers 2018, pp. 110, 45), Stengers questions the privileged standpoint of the critical academic as the authoritative voice of the problem of the Anthropocene. For Stengers, critique as remedy has become a poison, an end in itself for privileged academics that merely ratifies '*something that has already happened*' (2015, p. 111, emphasis in original). Reclaiming the concept of the pharmakon to address the question of climate disorder, as neither poison nor remedy but uncertainty, Stengers offers another reading of Swyngedouw's critique of ecology as the opium for the masses. Paying attention to the questioning situation of the Anthropocene and people's engagement with their milieu involves ' "thinking by the middle/milieu" ' (Stengers 2015, p. 103). Thinking in terms of the situations that cause people to think, feel, imagine, and act, rather than their fear, stupidity, or incapacity to think for themselves requires dismantling academic habits of critique and 'embracing the messiness of the world' (Savransky and Stengers 2018, p. 145). For Stengers, cultivating the art of paying attention to an ecology of practices, 'does not approach practices as they are' but 'as they may become' (2005, p. 186). Taking the ecology of practice as a tool for thinking is not a matter of reflexivity, argues Stengers, but rather demands an ' "ecology of partial connections", which requires learning from others, being transformed by what is learned, and acknowledging our debt to this transformative experience as we explore its problematising impacts in our own terms' (2018, p. 127).

In aiming to avert 'critical pondering from a safe distance', Stengers' (2021, p. 89) interest in modes of problematisation aligns with Beck's theory of metamorphosis, as an invitation for social theorists and social scientists to develop an imagination capable of grasping the transformative processes of thought happening in the world. By approaching Foucault's (1984) understanding of problematisation as a transformative engagement, that forces the thinker to test the limits of *our* thinking, Stengers privileges the power of a 'practical situation becoming a matter of explicit concern, a concern which cannot be met by epistemology or by the acquisition of (critical or positive) knowledge'. What matters are the apparatuses that gather concerned people in real time, around concrete situations, and the transformation of a practice into a cause for collective thinking. In reclaiming practices that experiment with the possibility of living in the ruins through Foucault's history of problematisation, Stengers is sceptical of critical analyses of climate disorder for failing to question how the uncertainty of the climate crisis might claim the power to make *us* think. Rather than disqualify what people could become capable of, problematisation demands engagement all the way down, 'involving partners for whom such practices matter, not choosy, off-ground, theory armed onlookers'. For Stengers, participating in the transformative power of questioning situations which force thinking becomes an ontological not philosophical achievement that is activated by a 'questioning situation that has already started to matter, even in an inchoate manner' (2021, pp. 79, 89, 80). Addressing transformative engagements that involve a 'metamorphosis of thought' (Savransky and Stengers 2018, p. 141) as problematisations awakens the imagination to the challenges of climate disorder for testing the limits of critical thinking and the contemporary reality of the academic to grasp the change it might induce in our own practice as thinkers and researchers (Savransky 2021).

The capacity to grasp the implications of Stengers' provocation for crafting new methodologies for thinking and research practices in the climate and plastics crises is addressed further in Mike Michael's (2017) work. In his analysis of how futures are enacted by social scientists, Michael takes inspiration from Stengers' concept of an ecology of practice. What interests Michael are the different ways the sociology of the future has been fashioned and enacted by social scientists in academic practice and how publics 'enact their own intellectual (and political Big Futures)' in mundane social and technological processes and practices differently. Comparing how Big Futures are enacted through social scientific analyses of Big changes, including those related to controversies of climate change, to how Big and Little Futures can emerge though uncontroversial mundane interactions with everyday technologies, including those related to mundane drinking practices, Michael makes reference to Shove's (2003) work, discussed earlier in this chapter. In Shove's analysis of everyday practice and the use of mundane technology such as the provision and supply of water, Michael points out she 'persuasively argues that these mundane technologies can be critically regarded as contributing to all manner of environmental problems'. In foregrounding mundane technologies in everyday practices as the lens through which sociology pursues an analysis of society, the mundane in Shove's work 'is represented as a mere illustration of a Big Future'. For Michael, the problem with Shove's sociological account of practice is that whilst a single conceptual framework is deployed to exemplify Big Futures, Shove's use of mundane technology to capture environmental problems is unable to address the unexpected side effects and potential of mundane interactions with everyday technologies to yield Big Futures. What's missing in Shove's analysis, according to Michael, is the question of 'how novel Big and Little futures can emerge out of mundane encounters' (2017, pp. 513, 516, 516, 511). The trouble with treating everyday embodied practices of drinking and the invisible technologies of water as a topic of critical analysis is 'the mundane can end up serving as an illustration of a particular critical version of the world'. In order for social scientists to avoid standing outside of these mundane events – with a 'particular analytic armoury in place' – and imagine how mundane technologies in everyday life might be collectively engaged with, Michael deploys mundane artefacts as *probes* to pursue this question empirically. Using probes as a means of engaging the mundane diverges from the methods of social science. Probes entail prompting engagement with the everyday that 'takes the user outside of the ordinary: they render the familiar unfamiliar'. For Michael, this enables participants to use probes in ways that 'generate responses that are unanticipated, creative, provocative, open and ambiguous' (2015, pp. 84, 84, 91, 90).

Outlining the difficult question of how to engage mundane practice and probe everyday futures, Michael (2017) contrasts Shove's study of social practice with Hawkins et al.'s study of plastic bottles and mobile drinking. Hawkins et al.'s empirical study of plastic bottled water practices is seen as an example that takes seriously the task of grasping the link between mundane interactions and everyday futures. Recognising the relevance of practice itself as a lens for thinking (otherwise) about the way mundane technologies can effect users' engagement in the world in unexpected ways is what Michael calls an ecology of futures. What interests Michael is how we 'attend to the potential for the Little Future of everyday technologies to *yield* (as opposed to *exemplify*) Big Futures' (p. 516, emphases in original).

By bringing together Shove's and Hawkins et al.'s studies as a way for thinking about how to open up possibilities of a different way of enacting the future in unforeseen Futures, Michael's analysis situates the future as a discrete sociological topic. Taking Stengers' question of why we should take seriously mundane practices as a tool for thinking about the power of a situation to matter, Michael shifts the focus of inquiry on global climate change from the study of everyday practices as social facts to the embodied knowledge and tacit skills that emerge through the process of practice. By situating knowledge of practice as emergent through social and technological processes (rather than a static category evaluated through a single conceptual theory of practice), Michael suggests Stengers offers another way of thinking about the enacting of futures in social scientific practices that does not privilege or take for granted our theoretical constructs and methodological practices, and is receptive to the futures enacted by other experts, skilled laypersons, and practitioners from other domains. What this means in practice, he argues, is an opportunity to engage with practical mundane engagements with the world and competing accounts of an event or object. For Michael, this involves posing questions in relation to the practices and techniques that have 'yielded and enacted this future' and 'opening up the possibilities for new con-joint practices' (p. 521) that are more modest about the claims we make, and more receptive to the futures enacted by colleagues.

Probing problematisation

The remainder of this chapter steps beyond Hawkins' investigation of the methods of anti-bottle activists and instead takes inspiration from Hawkins et al.'s (2015) description of the 'Do Something' water fountain campaign to yield Little and Big futures. This move presents an alternative account of problematisation to Hawkins' evaluation of the technique of 'moral problematisation' in environmental awareness campaigns, and 'pragmatic problematisation' in mundane practices, which acknowledges plastic materialities' capacity to 'trouble' the politics of environmental education (2010, p. 136). In seeking to probe problematisation in the 'Do Something' campaign, my intention is not to adjudicate between the deployment of problematisation as a technique of activists or academics but consider further what we can learn from re-engaging with Hawkins et al.'s (2015) thin descriptions of the mundane technology of the new public drinking fountain. Description enables us to pose questions in relation to mundane practices in problem spaces and shows how it becomes possible to respond to the methodological conundrum of engaging an ecology of practice in social scientific research that is receptive to futures enacted by others.

Following Nicole Vitellone, Michael Mair, and Ciara Kierans' (2020, p. 317) account of the promise of description 'to disrupt, unsettle and call into question disciplinary practices', my goal is to draw attention to what description offers for linking together mundane objects and everyday practice in the 'Do Something' water fountain campaign and what this opens up for investigating problematisation as a topic of inquiry. In their observation of 'Do Something' experiment, Hawkins et al. describe how the new public device of the water fountain 'changed relations between and among pedestrians, water, and drinking habits' (2015, p. 179). Their description of the new water fountain, as transforming and transformative of relations between human and nonhuman actors, connects the mundane technology of the fountain with everyday drinking practices and urban public settings. By switching the focus of their inquiry from the controversies and representational techniques of anti-bottle environmental activists to contest plastic water to a questioning situation, Hawkins et al.'s research offers another way of investigating the 'Do Something' campaign. By situating the water fountain 'as simultaneously the object, the site, and ultimately substance of thinking' (Rabinow 2002, p. 139), Hawkins et al.'s description of this mundane drinking technology as changing relations between and among pedestrians, water, and drinking habits tests the limits of thinking problematically and the form it takes. Describing the water fountain as producing new drinking practices calls into question the use of theory and methods and imagines new kinds of data. Engaging with the water fountain descriptively responds to the conundrum of grasping the ordinariness of problematisation that does not privilege theoretical constructs or taken-for-granted methodological practices.

As we have seen in previous chapters, the challenge of engaging problematisations as topics of inquiry in urban settings have been outlined by Clive Barnett (2015), Barnett and Gary Bridge (2017), Thomas Osborne and Nikolas Rose (1997), Osborne, Rose, and Mike Savage (2008), and Andrew Barry (2021). The pressing methodological task for social scientists requires a departure from deploying problematisation as a critical reflective practice and procedure to be followed by researchers and a turn towards more descriptive analyses of situational provocations. Arguing against the dominant interpretation of problematisation by social researchers and political theorists as the privileged subject of critical insight (Barnett 2015), and actor network theorists 'as a process of strategic interventions through which problems are defined', Barnett and Bridge (2017, p. 119) and Barnett's interest in urban situations of problematisation is of direct significance for making visible the 'Do Something' campaign as an example of problematisation that links mundane interactions and everyday futures. By shifting the focus of inquiry in Hawkins et al.'s study from the representational techniques of activists to a smaller scale questioning situation that provokes thinking, this chapter has proposed another way of researching environmental campaigns to contest the normalisation of plastic bottled water and reduce plastic water consumption. Using Hawkins et al.'s example of the 'Do Something' water fountain campaign as a case study to probe problematisation and how knowledge of plastic pollution is achieved on a smaller scale on the basis of everyday practice, the chapter has called for attentiveness towards concrete situations and the mundane technologies around which problematisations are achievable. Reimagining Foucault's notion of problematisation, as a situated, shared activity that concerns others engaged in a questioning situation in specific problem spaces, holds promise for investigating environmental campaigns to install new refill public water fountains. Paying attention to what is seen and describing what is not paid attention to in mundane settings suggests descriptive methods may provide a resource for developing affirmative modes of engaging metamorphosis in the everyday practices of others and our own research practices. This opens the more general question on *how* to conduct empirical research on the refill water fountain, which we approach in the next chapter through my study of London's #OneLess refill pilot. Through a consideration of interdisciplinary modes of inquiry, the chapter demonstrates the process of developing a research project on the #OneLess campaign and the challenges encountered in undertaking collaborative research with the academic and non-academic stakeholders and publics involved in #OneLess' refill water fountain experiment to reduce single-use plastics.

References

Badiou, A. (2008) 'Live Badiou – Interview with Alain Badiou. In O. Feltham [Ed.] *Alain Badiou – Live theory* (pp. 136–139). Continuum.

Barnett, C. (2015) On problematization: Elaborations on a theme in 'Late Foucault'. *Nonsite.* Issue 16, June 22.
Barnett, C. and Bridge, G. (2017) The situation of urban inquiry: Thinking problematically about the city. *International Journal of Urban and Regional Research* 40(6), 1186–1204.
Barry, A. (2021) What is an environmental problem? *Theory, Culture & Society* 38(2), 93–117.
Beck, U. (2015) Emancipatory catastrophism: What does it mean to climate change and risk society? *Current Sociology* 63(1), 75–88.
Beck, U. (2016) *The metamorphosis of the world.* Polity.
Foucault, M. (1984) Polemics, politics and problematizations. In P. Rabinow [Ed.] *The Foucault reader* (pp. 381–390). Pantheon Books.
Garfinkel, H. (1967/1984) *Studies in ethnomethodology.* Prentice-Hall.
Haraway, D. (1988) Situated knowledges: The science question in feminism and the privilege of partial connections. *Feminist Studies* 14(3), 575–599.
Hawkins, G. (2010) Plastic materialities. In B. Braun and S. Whatmore [Eds.] *Political matter: Technoscience, democracy and public life* (pp. 119–138). University of Minnesota Press.
Hawkins, G. (2011) Packaging water: Plastic bottles as market and public devices. *Economy and Society* 40(4), 534–552.
Hawkins, G. (2013) The performativity of food packaging: Market devices, waste crisis and recycling. *The Sociological Review* 69(S2), 66–83.
Hawkins, G. (2018) Qualifying. In C. Lury, R. Fensham, A. Heller-Nicholas, S. Lammes, A. Last, M. Michael and E. Uprichard [Eds.] *Routledge handbook of interdisciplinary research methods* (pp. 339–342). Routledge.
Hawkins, G. (2019) Mobile drinking: Mobile water practices and ontological politics. In C. Maller and Y. Strengers [Eds.] *Social practices and dynamic non-humans: Nature, materials and technologies* (pp. 109–128). Palgrave.
Hawkins, G., Potter, E. and Race, K. (2015) *Plastic water: The social and material life of bottled water.* MIT Press.
Kimbell, L., Guggenheim, M. and Marres, N. (2018) Inventive tensions: A conversation. In N. Marres, M. Guggenheim and A. Wilkie [Eds.] *Inventing the social* (pp. 317–333). Mattering Press.
Marres, N. (2012) Experiment: The experiment in living. In C. Lury and N. Wakeford [Eds.] *Inventive methods: The happening of the social* (pp. 76–96). Routledge.
Marres, N. (2015) *Material participation: Technology, the environment and everyday publics.* Palgrave.
Marres N., Guggenheim, M. and Wilkie, A. (2018) Introduction: From performance to inventing the social. In N. Marres, M. Guggenheim and A. Wilkie [Eds.] *Inventing the social* (pp. 17–35). Mattering Press.
Michael, M. (2015) Engaging the mundane: Complexity and speculation in everyday technoscience. In J. Chilvers and M. Kearnes [Eds.] *Remaking*

participation: Science, environment and emergent publics (pp. 81–98). Routledge.

Michael, M. (2017) Enacting Big Futures, Little Futures: Toward an ecology of futures. *The Sociological Review* 65(3), 509–524.

Mythen, G. (2018a) Exploring the theory of metamorphosis: In dialogue with Ulrich Beck. *Theory, Culture & Society* 35(7-8), 173–188.

Mythen, G. (2018b) The metamorphosis of the world: Society in pupation? *Theory, Culture & Society* 35(7-8), 189–204.

Mythen, G. (2020) Ulrich Beck: E-special introduction. *Theory, Culture & Society* 37(7-8), 383–409.

Osborne, T. and Rose, N. (1997) In the name of society, or three theses on the history of social thought. *History of the Human Sciences* 10(3), 87–104.

Osborne, T., Rose, N. and Savage, M. (2008) Reinscribing British sociology: Some critical reflections. *The Sociological Review* 56(4), 519–534.

Rabinow, P. (2002) Midst anthropology's problems. *Cultural Anthropology* 17(2), 135–149.

Savransky, M. (2021) Problems all the way down. *Theory, Culture & Society* 38(2), 3–23.

Savransky, M. and Stengers, I. (2018) Relearning the art of paying attention: A conversation. *SubStance* 47(1), 130–145.

Shove, E. (2003) *Comfort, cleanliness and convenience*. Berg.

Shove, E. (2010a) Beyond the ABC: Climate change policy and theories of social change. *Environment and Planning A* 42(6), 1273–1285.

Shove, E. (2010b) Social theory and climate change: Questions often, sometimes and not yet asked. *Theory, Culture & Society* 27(2-3), 277–288.

Shove, E. (2010c) Sociology in a changing climate. *Sociological Research Online* 15, 12.

Shove, E. (2012) Habits and their creatures. In A. Warde and D. Southerton [Eds.] *The habits of consumption* (pp. 100–112). Collegium.

Shove, E., Pantzar, M. and Watson, M. (2012) *The dynamics of social practice: Everyday life and how it changes*. Sage.

Shove, E. and Spurling, N. [Eds.] (2013) *Sustainable practices: Social theory and climate change*. Routledge.

Stengers, I. (2005) Introductory notes on an ecology of practices. *Cultural Studies Review* 11(1), 183–196.

Stengers, I. (2015) *In catastrophic times: Resisting the coming barbarism*. Open Humanities Press.

Stengers, I. (2018) *Another science is possible: A manifesto for slow science*. Polity.

Stengers, I. (2021) Putting problematisation to the test of our present. *Theory, Culture & Society* 38(2), 71–92.

Swyngedouw, E. (2010) Apocalypse forever? Post-political populism and the spectre of climate change. *Theory, Culture & Society* 27(2-3), 213–232.

Urry, J. (2011) *Climate change & society*. Polity.

Vitellone, N., Mair, M. and Kierans, C. (2020) Doing things with description: Practices, politics, and the art of attentiveness. *Qualitative Research* 21(3), 313–323.

Whitehead, A. (1968) *Modes of thought*. The Free Press.

3 Loitering as a method of interdisciplinary research

As we have seen in Chapters 1 and 2, recent developments in the environmental social sciences and science and technology studies (STS) do more than sound warnings about the plastics and climate crises. They also offer new ways of helping us make sense of how to respond to environmental problems. This has involved human geographers, social scientists, and STS scholars offering methodologies for engaging with environmental campaigns and experiments in more sustainable living. In the previous chapters, we have seen how situating and unfolding problematisation in the scholarly practices of human geographers, philosophers, social scientists, and STS-influenced scholars lead to new insights on how to investigate the mundane technology of the public water fountain and plastic bottled water. These insights have opened up a series of questions which probe everyday engagements with the water fountain in social research practices. For one thing, they raise questions about how these methodological insights can be brought into dialogue with other disciplines in the natural sciences.

In this chapter, we consider how London's #OneLess refill water fountain pilot opened a space for an encounter across the natural and social science. Focusing on my research encounters with marine biologists and conservationists who led the #OneLess campaign at the Zoological Society of London (ZSL), the chapter probes social scientists' capacity to engage with marine scientists and be challenged and transformed in the process. The other side of the question is how to acknowledge the contribution social scientists might make to environmental experiments. As we shall see, the conduct of interdisciplinary research provides an opportunity to examine social scientists' role in experiments in sustainable living and enables a consideration

DOI: 10.4324/9781003616757-4

of sociological methods and techniques as a topic of inquiry. In this sense, we consider not so much how environmental interdisciplinary research may be a potentially problem-solving intervention but how the #OneLess refill water fountain experiment brought into focus and made explicit the methodological, epistemological, and ontological assumptions that are often implicit in professional disciplinary practices.

Interdisciplinary research that moves scholarship from critical studies of problems in the world to thinking and collaborating with different disciplines, human and nonhuman actors, and stakeholders may seem obvious to those involved in studying environmental problems. However, as Andrew Barry and Georgina Born (2013) argue, we should not assume the methods of interdisciplinary environmental research are homogenous, uncontested, unified, coherent, or necessarily inventive and transformative. The diverse heterogeneous practices of interdisciplinary environmental research, they argue, involve a multitude of different social scientific approaches, techniques, and alternative perspectives that are marked by differences between not only the social and natural sciences but different social science perspectives. What the problematisation of interdisciplinary research seeks to provoke is a re-evaluation of interdisciplinarity as the basis for illuminating both the productive and anti-inventive dynamics within disciplines and the emergence of distinct interdisciplinary formations between the natural and social sciences (Barry and Born 2013).

To engage this task of addressing the methodological, epistemological, and ontological issues that arise when researchers from different disciplines encounter each other for the purpose collaboration, we begin by looking at contrasting analyses of interdisciplinarity in Barry and Born's (2013) text *Interdisciplinarity: Reconfigurations of the social and natural sciences*. Comparing the description of the inter discipline of geography by Sarah Whatmore (2013) with the inter discipline of sociology by Thomas Osborne (2013), we address the capacity of different methodological approaches and techniques to challenge and transform disciplinary and interdisciplinary knowledge practices. Outlining the research strategies and practices of sociologists and social scientists through Osborne's (2013) and Mike Michael's (2012, 2013, 2022) analyses of an idiotic methodology, the chapter illuminates the productive and anti-inventive dynamics of social research methods to inform knowledge of the #OneLess campaign. Using my own reflections on interviewing marine scientists, systems change practitioners, and publics involved in the #OneLess refill water fountain experiment, the chapter highlights the technique of interviewing as a knowledge practice that enabled the emergence of interdisciplinary research with natural and social scientists and informed new ways of investigating the #OneLess campaign to reduce single-use plastic water.

Interdisciplinarity as a problematisation

Reviewing the formation of interdisciplinarity in the social and natural sciences in the contemporary context, Barry and Born (2013) take Michel Foucault's concept of problematisation as the starting point to describe the present moment of knowledge production. What concerns Barry and Born are the polarised knowledge claims about the current status of interdisciplinarity as creative and fluid on the one hand, and disciplinary knowledge practices as 'repressive' (p. 7) and 'fixed' (p. 8) on the other. What troubles Barry and Born is the mistaken contrast between the 'restrictive structures' of disciplinary knowledge, and 'heterogeneity and openness of interdisciplinarity' (p. 7). Rather than assume the historical present is one that concerns such polarised binary oppositions, and a movement towards interdisciplinarity as a homogeneous shift in knowledge production, Barry and Born take up Foucault's notion of problematisation to inquire into the conduct of interdisciplinary research. What interests Barry and Born are the practices, processes, and events 'of what is potentially inventive or anti-inventive in the emergence of interdisciplinarity in particular eras and fields' (p. 6).

In their problematisation of the practices that constitute interdisciplinarity as an object of thought in scientific research, Barry and Born probe the diverse modes and logics of interdisciplinary practices and forms of knowledge production. In the context of climate change and interdisciplinary environmental research, they find a widespread taken-for-granted view of the social scientist as enacting an '*integrative synthesis* mode' of interdisciplinarity, through a process of integration or negotiation. This ideal typical mode of practice, they argue, 'should not necessarily be understood additively as the sum of two or more "disciplinary" components' (p. 11, emphasis in original). Instead, they point out, the social scientist often takes a '*subordination-service* mode' of interdisciplinary practice, where their role is 'typically conceived as making up for, or filling in for, an absence or lack in the other, (master) discipline(s)' (p. 11, emphasis in original). In drawing attention to the hierarchical division of labour in interdisciplinary research, Barry and Born propose a contrasting '*agnostic-antagonistic*' mode of interdisciplinarity, as involving a particular problematisation, which we explore in more detail in this chapter (p. 12, emphasis in original). What distinguishes the agnostic-antagonistic mode of interdisciplinarity is neither a synthesis of two separate disciplines, nor a hierarchical disciplinary division of labour, but a radical shift in knowledge practices. This involves a commitment to 'contest or transcend the given epistemological and/or ontological assumptions of specific historical disciplines, a move that makes the new interdiscipline irreducible to its "antecedent disciplines" ' (p. 12). The practice of interdisciplinarity as problematisation, for Barry and Born, entails the academic researcher engage in a 'conscious dialogue with, criticism of or opposition to the limits of established disciplines' (p. 12). Within interdisciplinary environmental research, Barry and Born point out this agnostic-antagonistic mode of problematisation is governed by an ontological logic that takes as its focus efforts to transform research practice. Radical transformative practices of inter-disciplinary environmental knowledge involve; intervening in the world by 'bringing the subjects and objects of research into a relational existence'; generating 'relational objects that cannot be broken down into distinct natural, technical and social components' (p. 19); and recognising the limitations of scientific expertise and professional analyses of global environmental problems by including non-experts and publics in scientific research, in order 'to generate both different objects and different ways of proposing problems' (p. 20).

Reflecting on the tensions between two interdisciplinary imperatives at play in geographical practice, Whatmore (2013) draws attention to the risk and potential of her own discipline of geography to be both inventive and anti-inventive in producing transformative interdisciplinary knowledge. Discussing her professional experience as a human geographer, Whatmore outlines the role of the geographer in institutional environmental and climate change research. On the one hand, Whatmore highlights the normative role of enacting an integrative-synthesis model of interdisciplinarity, in which research is 'positively allied to governmental and business agendas' (p.163). On the other, Whatmore questions the 'intellectual cost' (p. 162) of a subordinate service model of expertise, in which interdisciplinary energies are aligned with institutionalised power. Sceptical of geographers embracing the role of 'environmental problem solving', Whatmore (p. 163) describes how collaborative experiments between the natural and social sciences create an agnostic-antagonistic mode of interdisciplinarity, in which knowledge practices are negotiated. For Whatmore, the privileged, ambivalent position of geography in interdisciplinary research collaborations enables the discipline to resist the integrative mode of negotiating knowledge from a subordinate position. Geography, argues Whatmore, has the potential to engage in the problematisation of the conduct of research, in which the production and contestation of environmental knowledge practices of different disciplines and stakeholders is foregrounded.

What's necessary for this transformation in interdisciplinary knowledge practices is a switch in focus from technical questions of how to solve environmental problems to 'an ontological question of de/reconstructing their assemblage' (p. 171). In drawing attention to the logic of ontology in interdisciplinary problematisation – as enabling the contestation and transformation of research practices – Whatmore cautions that there is no 'single, or ready-made formula for realising the potential of geography's interdisciplinary inventiveness' (p. 173). At the same time, Whatmore warns against relying on the deconstructive practice of problematising the professional conduct of environmental problems. Critiquing institutionalised power alone, argues Whatmore, 'is not sufficient to either the political or scientific demands of such problems' (p. 174). Geographies' unique disposition in interdisciplinary problematisation involves the possibility of '*re*constructing how environmental research is done' (p. 174, emphasis in original). By paying attention to 'the artefacts and processes assembled in/as "environmental problems"' (p. 163), and the experience of non-experts about what is at issue, Whatmore extends knowledge practices in the agnostic-antagonistic model of environmental interdisciplinarity to intermediary objects and the local sites of experimental practices of problematisation. What this means in practice is 'invent[ing] apparatuses in which those whose experiences makes them sensible of and knowledgeable about *what* is at issue are enjoined in the exercise of more effective public environmental reasoning' (p. 174, emphasis in original).

The *where* of interdisciplinarity is also foregrounded by Osborne (2013) in his contribution to Barry and Born's edited collection on the problematisation of disciplinary and interdisciplinary knowledge practices. What concerns Osborne is the conduct of social science disciplines and their exaggerated claims to interdisciplinarity. Whilst he acknowledges different disciplines have different ways of knowing and interacting with other disciplines, Osborne is sceptical that different modes of interaction be understood as interdisciplinary problematisations, or that understanding interdisciplinarity as an integrative synthesis of two separate fields of inquiry is the opposite of disciplinarity. The dichotomy of disciplinarity and interdisciplinarity, argues Osborne, 'is completely misleading' (p. 82). Interactive tendencies in the social sciences for Osborne are integral to the conduct not of interdisciplinary but disciplinary practices. Rather than being opposed to disciplinarity, Osborne argues interdisciplinarity 'assumes a certain consciousness of disciplinarity as a condition for its accomplishment' (p. 82). Comparing the capacity of the disciplines of anthropology, economics, and sociology to address different modes of (inter)disciplinary problematisations, Osborne describes the creative and negative status of social scientific knowledge practices in terms of varying degrees of 'disciplinary porosity' and 'promiscuity – the different ways they get themselves about' (p. 86). What interests Osborne is the logic of parasitism, trespassing, and poaching in different modes of disciplinary interaction in the social sciences.

Describing the porosity of anthropology in a creative sense, Osborne highlights the ethnographic method of participant observation as providing anthropologists with unique capabilities and techniques to go out into the field, ready 'to be absorbed into it' (p. 86). The parasitic quality of anthropologies methodological practice, argues Osborne, enables anthropology to 'exist as parasites on other domains', 'while leaving their own disciplinary core intact' (p. 87). In his evaluation of interdisciplinary problematisation in the conduct of anthropological research, Osborne suggests the porosity of anthropologists ethnographic practice is a 'part of their style of being *disciplinary*, not something to do with interdisciplinarity as such' (p. 86, emphasis in original). However, Osborne is less affirmative in his assessment of interdisciplinarity in the field of economics. Describing the example of the Stern Report (2006) on the economics of climate change as an example of 'impos[ing] your own view of the terrain on another disciplinary area' (p. 88), Osborne describes economics as having 'a magnificent propensity for trespassing' (p. 89). The trespasser does not so much take from other domains of environmental research, but 'passes into other domains and develops his or her own perspective' (p. 88) *without* any interaction with other disciplines. Although Osborne calls into question the practice of trespassing for its reductive rather than interdisciplinary tendencies, his greatest criticism concerning interdisciplinary knowledge is levelled at the academic practice of poaching. Describing poaching in the social sciences as a reverse form of trespassing, Osborne argues the poacher enters another field, takes what they understand to be the key insights of that field, then 'seeks to return these to his or her own area of research' (p. 90). The problem with poaching, explains Osborne, is not only that it is not interdisciplinary but that poaching 'isn't disciplinary enough', symptomatic of a 'disciplinary deficit' that is undertaken by someone who doesn't know what they're doing in their own discipline (p. 90).

Osborne's evaluation of the ontological logic of porosity and promiscuity in social science disciplines highlights the strengths, limitations, and failures of interdisciplinary research as a practice of disciplinary reformation. At the same time, Osborne argues interdisciplinarity is the most exciting 'where the creative energies of its practitioners are most at stake in entering into the unknown' (p. 95). The novelty of disciplines harmoniously 'working alongside' (p. 96) each other, in a platonic rather than an agnostic-antagonistic mode of inquiry, argues Osborne, requires collaborating, communicating, evolving, and synergising as disciplines. The summative meeting of two or more disciplines 'learning to work together and to pool results – … platonically laying them alongside each other in relation to the overall outcome' (p. 92) is more likely to produce an interrogation of one's own disciplinary practices and 'destabilise disciplines in the sense of provoking new scales and ranges of objects and problems for them to confront' (p. 92). In contrast to the discipline of anthropology, which is understood as well suited to the task of working alongside other disciplines, Osborne points out sociology is considered less adept at engaging in a mode of platonic interdisciplinarity. The difference, according to Osborne, is not that the method of ethnographic fieldwork is more interdisciplinary but that sociology by contrast is 'loosely disciplined' (p. 96) and less distinctive in terms of its methods. Without a distinct methodology or object of research, sociology 'can't make much headway in the interdisciplinary stakes' (p. 96). Whilst sociologies' lack of stable fixed methods, conventions, and consensus of what it is for or against may be seen as a bad thing, this lack of a disciplinary identity, argues Osborne, is where sociologies strengths lie. In the present situation, dominated by problematisations of disciplinarity and interdisciplinarity, sociology's capacity for 'scepticism', 'stubbornness', and 'creative idiocy' 'open up a space *to think*' (p. 96, my emphasis).

Crafting an idiotic methodology

Osborne's assessment of sociologies contribution to interdisciplinary problematisation as an '*idiot science*' (p. 96, emphasis in original) adapted from Gilles Deleuze and Felix Guattari's (1994) figure of the idiot is more broadly explored by Mike Michael (2012, 2013, 2022). Responding to Barry and Born's (2013) problematisation of interdisciplinary collaborations between divergent disciplines, Michael (2013, 2022) proposes a methodology for social scientists in interdisciplinary encounters as neither agnostic-antagonistic nor platonic, but idiotic. Drawing on Isabelle Stengers' definition of Deleuze's figure of the idiot, as that which 'resists the consensual way in which the situation is presented and in which emergencies mobilize thought and action' (Stengers 2005, cited in Michael 2013, p. 73), Michael describes the nonsensical idiot as a conceptual persona that enables the opening up of encounters with disciplinary others and the slowing down of thought. For Stengers (2005), the idiot demands that we slow down thought and 'don't consider ourselves authorized to believe we possess the meaning of what we know' (p. 995) through the notion of theory. The idiot creates 'an opportunity to arouse a slightly different awareness of the problems and situations mobilizing us' (p. 994), a knowledge practice Stengers describes as a cosmopolitan proposal. What characterises the ontology of the idiot in Stengers cosmopolitan proposal is the expert's vulnerability of not knowing *how* to proceed. In privileging Stengers' (2005) definition of the idiot as troubling conventional disciplinary and interdisciplinary knowledge practices, Michael's reflections on the uses of the idiot in social science expands Osborne's description of sociology as an idiotic science. In an idiotic encounter, Michael explains 'practitioners of one discipline can simply not grasp – find nonsensical – the objects of study of the other discipline' (2013, p. 79). The mutual idiocies of divergent disciplines are understood not as an interdisciplinary problematisation to be negotiated, contested, or transcended through conscious dialogue and critique but a live problematisation to be actively engaged in by social scientists through the practice of research.

Unlike the practices of parasitism, trespassing, and poaching, which Osborne (2013) describes as operating through a model of disciplinary surplus or deficit, requiring little interaction with other disciplines, Michael (2013, 2018, 2022) highlights the potential uses of Stengers' cosmopolitan proposal for the social sciences in encounters with disciplinary others. This involves different disciplines engaging in the ontological logic of compromising. The potential of interdisciplinarity to 'trouble the conventional and challenge the consensual', argues Michael (2013, p. 77), relies on the presence of the conceptual persona of the disciplinary idiot to 'facilitate a slowing down (not to jump too hastily to the view of others' viewpoints and practices)' (2022, p. 108) and provoke thought. The aim is to acknowledge the views and practices of disciplinary others and accommodate their differences and similarities without 'mov[ing] to a '"common ground" too easily' (2018, p. 281). What compromising means in practice, Michael explains, is 'paying attention to that which does not make obvious sense within an unfolding collaboration (the idiotic in Stengers' terms)' (2018, p. 281). In order not to impose one's disciplinary baggage, or arrive at compromise too easily or without compromise, Michael (2018) suggests collaborators' nonsensical views and practices must not be denied through a positive alignment with the others disciplines or deferred to through a position of authority or subordination but actively acknowledged, accommodated, and valued. Engaging Stengers' (2005) practice of compromise in interdisciplinary research, Michael (2018) points out, is nothing less than a transforming and transformative task requiring great efforts of staging interactions with different disciplines and professional experts from a position of equality and openness that is sensitive to others' views and different practices about the issue, and how disciplines deal with it. It is also, as Stengers (2005) describes a vulnerable proposal. Achieving equality through compromise means making present both the role of experts in professional knowledge practices and consulting with invisible others and the murmurings of the idiot to slow down and prevent a shared consensus of what counts and what doesn't. While embracing the role of the disciplinary idiot in interdisciplinary problematisations has the potential to trouble commonsensical thoughts, and generate a different awareness of problems, objects of study, subjects, and relations, Michael's (2012, 2013, 2018, 2022; Wilkie and Michael 2023) methodological engagement with the conceptual persona of the idiot highlights the impact of the cosmopolitan proposal for slowing down dialogue across disciplines and resisting the consensual thinking of practitioners. It also highlights the uses of disciplinary incommensurability as a resource to draw out insights for, and about, different disciplines in the conduct of research that does not involve finding a common ground but sitting in an uncomfortable space.

In order to stay with the trouble of 'non-sense-making' provoked by our encounters with others in the research event, Michael (2022, p. 74) proposes three ways of disturbing the idiot – accidentally and deliberately, which we address in the paragraphs to follow and purposefully which we return to in the next chapter in my direct observations of the refill water fountain pilot. What we address here are accidentally encountered idiots in compromising situations I experienced in conducting interviews with marine scientists, professionals, and public participants involved in the #OneLess refill campaign. Being confronted by accidentally encountered idiots in conversations with experts from other disciplines, for Michael, enables social scientists to stop making professional sense of a situation and question what we think we are doing. Acknowledging incomprehension in compromising interactions with research collaborators disrupts and opens a space to 'slow down our thinking and reflect on our intellectual presuppositions' about the research event (p. 74). Paying attention to accidental encounters with research participants who do idiot things in the research event, and act in ways that don't make sense to the researcher, has practical implications for social scientific research. Foregrounding the actions and reactions of participants within the research process as idiotic forces us to slow down our thinking and question our 'researcherly presuppositions' (p. 71). Rather than ignore, exclude, or attempt to manage misbehaving others in our research interactions, Michael encourages us to embrace the accidentally encountered idiot to enable a rethinking of the conceptualisation of a research field. Becoming sensitive to nonsensical behaviour within an interview or participant observation is seen to put into question our existing disciplinary expertise, where everything about our research methods and professional practices makes theoretical and methodological sense. In the idiotic encounter, the meaning of the research event for the researcher is not shared by research participants. Acknowledging 'the potential mutuality of idiocy' (2012, p. 179) between the researcher and participant and what doesn't make sense challenges the research event as 'a site of "consensus"' (2022, p. 70), and prompts the asking of questions about what those research events mean.

In outlining that which enables social scientists to slow down their thinking and reflect on what is being done, Michael (2012) draws attention to a third role of the idiot for changing sociology and its methodologies. Rather than limit observation to misbehaving humans in the research event, he addresses accidentally encountered idiotic *objects*. In order to become sensitised to the role nonhumans play in slowing down thought and providing a pause to think about the ontological status of objects in society and sociology, Michael highlights the everyday objects encountered in the conduct of empirical study. In particular, he acknowledges the contribution of digital recorders, pen and pencil, tape recorders, and computers as tools in the process of making sociological research events such as interviews and ethnographies. What interests Michael is the idiotic role of these objects in the process of conducting social science and the making of social data. Reflecting on the example of a disastrous interview example where the tape recorder misbehaves, Michael probes the ontology of everyday objects in events that go wrong as an opportunity to slow down and reflect on the interview event and question our assumptions about the meaning of social scientific research and the gathering of data. By analysing the idiotic tape recorder that does things it isn't supposed to do, Michael (2004) prompts sociologists to take seriously the role of technological nonhumans in research practices. Reviewing the disastrous interview episode facilitates a rethinking into not just who, but what 'can take up the role of the "idiot"' (2022, p. 70). By attending to the nonsensical interview, Michael produces an alternative meaning of the interview event as idiotic, and 'reminds us that we must never get too comfortable with "what are we busy doing"' (2011, pp. 536–537).

Problematic interviews

Similarly, Les Back's (2012) broken tape recorder that fails to work during a two-hour interview stimulates a provocation that opens up possibilities to pause and think about the technical practice of interviewing and what sociologists are busy doing. In Back's reflections of his 'dead tape recorder' (p. 257), the idiotic object becomes a device of disciplinary problematisation. The accidental encounter with a misbehaving voice recording device enables a reflection of the researcherly suppositions of qualitative research and a rethinking of the conceptualisation of sociologies methodological imagination. In reviewing the failed interview recording episode, Back acknowledges the taken-for-granted role of the tape recorder as a professional instrument in the craft of doing sociology and questions whether the object has become an outmoded sociological device. While the use of the tape recorder distinguishes the discipline of sociology from anthropologies' methodological practice of fieldwork, and the ethnographic method of participant observation and description with its tools of pen and paper, Back's reflections on his broken tape recorder produces an interrogation of the idiotic disciplinary practice of interviewing. By slowing down thinking, the misbehaving object provides pause to reflect on the idiotic tape recorder as both an enabling and limiting sociological device. On the one hand, Back realises his dead tape recorder has been a 'companion in sociological sociability' and a 'sociological shield' (p. 251), which enabled him to be connected to the social world through the interview encounter and protect him when he felt at risk. On the other, Back acknowledges a sociologists' dependence and addiction to their tape recorder limits other forms of empirical documentation, such as observation and description. Not wanting to abandon the interview, and 'consign the tape recorder to the junkyard of outmoded devices' (p. 250), Back calls for the revitalisation and transformation of sociologies methodological techniques through the development of descriptive methods and devices that use sound beyond simply recording human voices.

Back's disciplinary problematisation of the professional practice, process, and objects of interviewing highlight what is potentially inventive and anti-inventive in the formation and reformation of the discipline of sociology. For Back, the historical present concerns a polarised view of the tape recorder and process of interviewing as setting boundaries and limits to sociological knowledge practices (see also Savage and Burrows 2007). By interrogating the attachment to his broken machine, Back calls into question taken-for-granted norms of the craft to block the sociological imagination. Acknowledging a nostalgic attachment to his tape recorder, Back contests the methodological assumptions of the discipline to restrict social scientific expertise and analysis. Letting go of the tape recorder is perceived as transcending knowledge practices by unblocking the uses of sound recording devices to capture the noises and rhythm of life as data. In Back's analysis, he thus foregrounds the advantages and disadvantages of the tape recorder as a distinctive sociological device for imagining and reimagining disciplinary knowledge practices. In the reflections on using the technique of interviewing to follow, I consider what attentiveness to the device of the audio recorder in the conduct of environmental interdisciplinary research enables and inhibits. Reflecting on my practice of interviewing marine scientists and conservationists, non-academic stakeholders involved in the #OneLess refill pilot, and the publics engaged in the use of the refill water fountain, we address the ontology of the audio recorder not as an anti-inventive device 'that has had its day' (Back 2012, p. 246) but an idiotic object for probing new ways of imagining disciplinary and interdisciplinary research practices. Recognising the materials with which we do our work are not neutral, invisible, or ignorable requires explorations that take us beyond specific tools into new theoretical territory (Paulus, Jackson and Davidson 2017). By making the devices and practice of interviewing the focus of my discussion, we address research methods not as tools for measuring problems and solutions to real world environmental issues but materials for reimagining how Foucault's concept of problematisation transforms knowledge making practices.

Reflecting on the relevance of Back's broken tape recorder at hand, Celia Lury and Nina Wakeford (2012) draw attention to the concept of device for thinking about the inventiveness of methods to address a specific problem and what emerges in the use of methods to change it. The concept of the device enables them to think about how the practice of interviewing and object of the tape recorder are linked to each other, with potentially inventive – or anti-inventive – effects. In thinking about the capacity of device, Lury and Wakeford link Back's reflection on the tape recorder to Foucault's notion of the dispositif and apparatus, 'as a system of relations' (Foucault 1980, p. 194, cited in Lury and Wakeford 2012, p. 8) and Agamben's definition of Foucauldian apparatus', 'as literally anything that has in some way the capacity to capture, orient, determine, intercept, model, control, or secure the gestures, behaviours, opinions, or discourses of living beings' (Agamben 2009, p. 14, cited in Lury and Wakeford 2012, p. 8). By locating the device in relation to Foucault's notion of dispositif and apparatus, Lury and Wakeford stress 'a device or method is never able to operate in isolation since it is always in relations that are themselves always being reconfigured' (p. 8). In linking the tape recorder and site of the interview as an apparatus for thinking about relational processes of knowledge production, Back's tape recorder is seen as acting as a hinge between concepts and practice, 'opening a door …

on to the practical investigation of the social world' (p. 9). Lury and Wakeford's inventory of devices thus makes room for thinking about Back's encounter with the dead tape recorder as a normal trouble of doing social research. The process of dealing with this trouble presents us with a lens on how the method of interviewing and device of tape recorder are linked to each other in practices of knowledge making, and with what effects. It also invites a reflection on how interviewing might contribute to interdisciplinary research and the role of interdisciplinary methods 'in the determination of a situation as a problem' (Lury 2018, p. 3).

While the deployment of Foucault's notion of dispositif and apparatus asks us to think about the capacities of the tape recorder as a hinge between concepts and practice, Barry and Born's (2013) engagement with Foucault's work on problematisation takes the object of the tape recorder in a new direction. While the recording device and method of interviewing have been problematised by academic experts, Barry and Born's engagement with Foucault's concept of problematisation goes beyond understandings of the tape recorder as entailing a conscious reflection on the capacity and limits of disciplinary practices and entails an engagement with experts from other disciplines, non-experts, and lay publics in interdisciplinary formations that may lead to the production of new objects, subjects, and relations of knowledge. In considering what the audio recorder contributes to disciplinary and interdisciplinary problematisations, I am not suggesting it is a pre-existing object for thinking about the method of interviewing independently of those practices (see Benney and Hughes 1956; Atkinson and Silverman 1997; Silverman 2007). Nor am I interested in problematising the use of disciplinary techniques of interviewing in qualitative research (see Rapley 2004). While the recording of interviews has been taken seriously as a practical matter of concern in the discipline of sociology, Raymond Lee (2004) argues there has been little attempt to develop a reflexive awareness of how recording methods affect the research process and how researchers and interviewees negotiate recording practices in the conduct of the interview. In seeking to address this gap, I take seriously Charles Briggs' (2007) call for social scientists and especially anthropologists to focus attention on interviewing practices in their research projects as an object of inquiry. In doing so, I attend to the use of audio recorder in the context of my own practice of interviewing as a material and social device that has multiple effects in the making and doing of disciplinary and interdisciplinary knowledge practices (Law and Ruppert 2013). Responding to Susan Nordstrom's (2015) invitation for scholars to stop taking their recording devices for granted, and make them matter differently through the use of theoretical tools, in what follows we consider how Foucault's notion of problematisation opens up a space to think with the audio recorder in the conduct of research, and the normal troubles encountered in its use. By investigating the audio recorder within the conduct of research interviews, we address how Foucault's concept of problematisation 'lives and breathes in the recording device' (Nordstrom 2015, p. 399). In order to grasp what a turn to problematisation makes possible, we examine a series of accidental encounters with interview subjects and the audio recorder. In attending to the social life of the audio recorder, we address the troubles encountered in the practical accomplishment of interviewing. Dealing with problematic interviews in local practices as a phenomenon to be studied for Christian Grieffenhagen, Michael Mair, and Wes Sharrock (2015) is an alternative starting point for addressing social scientists' investigative practices. The following examples aim to illustrate the role of technological nonhuman in the making of data in ways that reconfigure understandings of problematisation in disciplinary and interdisciplinary knowledge practices.

Compromising situations in researching the #OneLess campaign

In October 2018 I set off for London with my suitcase and audio recorder to start a new empirical research project on #OneLess' refill water fountain pilot which had recently been introduced into the capital. The period leading up to my departure was occupied with the normal professional business of getting ethics approval through the University of Liverpool central ethics committee to enable the conduct of qualitative research. The research proposal included interviews with key academic and non-academic stakeholders involved in the #OneLess campaign, observation of the refill water fountain pilot in use, and interviews asking the public to tell me about how they encounter and use the refill fountain in situ. The ethics application was heavily scrutinised and approved. Following a formal interview with the ethics panel, and subsequent minor amendments to the application, I was granted permission to undertake empirical research on the refill water fountain, following four conditions. First, I do not make myself conspicuous in public places while conducting fieldwork. Second,

I do not take photos of people when documenting my observations of the refill water fountain. Third, I secure written consent prior to conducting all interviews with professionals involved in the pilot and members of the public engaged in the use of the refill water fountain. And fourth, I gain consent from the relevant stakeholders to enter into the public spaces where I sought to undertake research. The process of requesting permission for access to three field sites across central London was conducted via email and phone calls. The nature of my sociological research was explained in a lay language and the content of my semi structured questions with the public shared. In the three locations where I requested permission to undertake research, access was granted. In all three sites, I was required to present myself to members of the security staff on arrival and departure of each day of fieldwork. At one of these sites, a digital photo of my face was taken on each visit and a day pass issued for me to wear at all times around my neck while on the premises (Figure 3.1). The plastic encased pass included my photo, the date, my name, and institutional details. These gestures of reporting to security staff, and displaying my institutional credentials, were seen as necessary processes to authorise the conduct

Figure 3.1 Disciplining Research.

of sociological research and legitimise my professional disposition in the field.

Whilst the practice of undertaking social scientific research is often taken for granted as having followed a straightforward trajectory, from research design to ethics approval, to access, and the conduct of research, what enabled the practice of observation and interviewing were the materials of my badge, participatory information sheets, consent forms, and a functioning audio recorder. These material devices had different idiotic effects that troubled the doing of sociological research and gathering of data. In Michael's (2012, 2022) account of the links between the idiot and the discipline of sociology, he draws attention to the misbehaviour of idiots encountered in interactions with professional colleagues from other disciplines and research participants. Rather than mistake the idiot for non-data, Michael (2012, 2022) asks us to look again at the nonsensical as something relevant that happened, and reflect on misbehaving objects and participants impartially, instead of ignoring them. While difficult to reconcile the idiot with the conduct of qualitative research, staying with the trouble of the problematic interview generates a different awareness of social research methods, objects, and data. In what follows, I describe a series of accidental idiotic encounters with disciplinary others, professionals, and public participants that generated a different awareness of my research practices.

My initial email requests to interview key stakeholders were responded to very quickly and positively, and I was invited to attend a meeting at the London Zoo with the #OneLess project team of marine scientists and systems change pracitioners involved in the pilot to discuss my research project. In my haste, I arrived a day early with my audio recorder, participant information sheets, and consent forms and returned the following day to conduct the first of a series of 6 unstructured interviews of approximately 1–2 hours with various members of the #OneLess team. There was nothing exceptional about the conduct of these face-to-face interviews, which involved information gathering about the #OneLess project, information sharing about my methods, and reporting my observational and interview findings with the public. But these unexceptional interactions became an important aspect of interdisciplinarity that provoked methodological reflections. All the interviews were conducted at the London Zoo, where I was met by a member of the #OneLess project team at a designated building entrance to gain access. The interviews took place in a quiet, accessible meeting room where I was seated at a table and able to speak and listen to others without interruption or disruption. During all the interviews I was offered a cup of tea. In many ways, my encounters with the #OneLess team resemble the platonic mode of inquiry described by Osborne (2013), where social scientists and natural scientists work alongside each other in a collaborative, communicative mode of interdisciplinary knowledge production. Yet, the method of interviewing opened up a space to think differently about the conceptualisation of interdisciplinarity. Despite the absence of strange encounters with misbehaving colleagues or idiotic objects, during one of my visits to ZSL I began an interview without seeking written consent. Although I'd packed the printed consent form and a pen, I forgot to ask for a signature before turning on the play button of my audio recorder. My professional misbehaviour was noticed and drawn to my attention by my interviewee shortly after the interview had started. I immediately pressed pause on the audio recording device and retrieved a consent form and pen from my bag which was signed. Whilst forgetting the consent form has the potential to render nonsensical the whole activity of the research interview if not dealt with, thankfully, I was able to repair my mistake without it disrupting the event at hand. Embarrassed by my error, and annoyed for interrupting the flow of the conversation, what also struck me was not just that I had forgotten to present the consent form, but that *they* had noticed. My idiotic behaviour of slowing down the interview enabled an opportunity to pause and reflect on the presence of the audio recorder as a taken for granted sociological device to capture voices. In calling attention to the idiotic practice of asking for signed consent before taking part in a recorded interview, my interviewee draws attention to the presence of the audio recorder as a device used not just for the purpose of data gathering and analysis but a relational object central to the process of engaging in interdisciplinary interactions.

What my misbehaviour revealed was an object whose ontology made possible a meeting with other disciplines. What was at stake in the accidental idiotic encounter between the printed consent form, pen, audio recorder, sociologist, and marine scientist was the research event. The audio recorder brought social and natural scientists together into a relational conversation. The interaction was enabled by an apparatus that produced compromise. By accidentally encountering a compromising situation with a research collaborator, my idiotic behaviour produced a questioning of disciplinary practices and reflection on the role of objects in interdisciplinary encounters. The accidentally encountered idiotic audio recorder and consent form gave us pause to think. In opening a space to think about interdisciplinary interactions, the paused audio recorder reveals what did not make obvious sense. By enabling the opening of an encounter with disciplinary others and a slowing down of the research event, the idiotic audio recorder engaged the sociologist and marine scientist in the practice of compromising. In requesting to sign a consent form, my interviewee actively acknowledges the practical doings of qualitative research and accommodates our methodological differences. In being forced to think about the idiotic logic of social scientific research, my attachment to the audio recorder loses its ontological grip as a sociological device to collect, transcribe, and re-circulate voices (Back 2012) and becomes a device central to the staging of a compromising interaction with a different discipline and professional expert.

The problematisation of the interview took a different turn in my encounter with a non-academic professional involved in London's refill experiment. In organising an interview with a key stakeholder not directly associated with the #OneLess campaign, I was asked to email my questions in advance of our in-person meeting. On the day of the research encounter, which took place in an office near the location of the water fountain, my request for consent to record the interview was denied, and the interviewee avoided answering my emailed questions. While this unwillingness to be interviewed presented practical implications for the doing of social scientific research, the refusal to partake in a recorded interview slowed down what sociologists busily do and brought to light researcherly pre-suppositions about the ontological status of objects in staging an encounter with non-academic others. The problem of being left without a signed consent form, and an inability to turn on my recording device, revealed taken-for-granted assumptions about the presence of the audio recorder in the conduct of a research interview. In calling into question the use of my research device, the meaning of the audio recorder in the interaction is not shared, and the research event as a site of consensus is challenged. Suspicion surrounding the device and practice of interviewing creates an antagonistic power imbalance that undermines the logic of compromise. By associating my recording device with the practice of taking and misappropriating the words of another, the method of interviewing resembles the ontological mode of poaching, trespassing, and parasitism. In problematising disciplinarity, the interviewee draws attention to the role of the audio recorder, consent form, and interview method as a hierarchical knowledge practice in sociological research. Calling attention to the non-sensical behaviour of a social scientist seeking consent to use their audio recorder, the misbehaving interviewee provokes thought about sociology's idiotic methodology. In challenging the conduct of qualitative research, the idiotic interview encounter creates an opportunity to arouse a different awareness of the presence of the recording device in the research event. Being forced to embrace the role of the idiot within the problematic interview provoked a rethinking of my disciplinary methods and professional practice.

Loitering as an interdisciplinary method

My accidental encounters with research participants in the field also drew attention to the ontological stakes of engaging the idiot in the practice of interviewing publics. Armed with my consent forms in a clipboard in one hand, my audio recorder in the other, and the badge identifying my professional status around my neck, my idiotic presence in urban spaces in three central London sites incited a range of curious and incurious actions and reactions from the public. These responses provoked a slowing down of thought and reconceptualisation of my disciplinary practice. Responses to my presence in these problem spaces included incurious members of the public asking me for information on where the toilets were, if I was their tour guide, and the police asking me to act as the 'eyes and ears' of terrorism and report any left luggage and strange behaviour. While these encounters drew attention to my idiotic attempt to create a field of research in the public setting of the refill water fountain, the curious response from one participant prompted a mutual idiocy that did not make me feel like my research efforts had gone to waste. On consenting to be interviewed, the research participant commented they had been watching my behaviour and could not make sense of why I was standing near the water fountain, or grasp what I was doing there. My idiotic behaviour of 'hanging around' the water fountain with my clipboard and audio recorder in hand, had been noticed, and my idle presence in a public place was described as '*loitering*'. The respondent's reflections on the ontology of loitering prompted further questions about what I was doing, why I was there, my research methods, and professional disciplinary practice.

While loitering might not be what social scientists claim they are doing when engaged in the process of interviewing and observation, the ontology of loitering in a fixed urban place for a period of time with no specific purpose or reason gives us pause to think. The idiotic strategy of loitering produced interactions with others that transformed understandings of disciplinary practices. In contrast to anthropologies parasitic tendencies, loitering provoked an accidental idiotic encounter of nonsense making that created a different awareness of the situation. My idle presence in a public space is not dismissed as unimportant, or treated as suspicious, but perceived as incomprehensible. The research participant's inability to grasp what I was doing, or why I was there, unsettles assumptions about what sociologists think they are busy doing. Loitering provokes an accidental idiotic encounter of nonsense making, that gives the social scientist and the research participant pause to think. By slowing down thinking, the ontology of loitering reveals the contribution of social research methods for the study of environmental knowledge campaigns. Investigating the #OneLess refill experiment is produced not by collecting data from the public and aligning with disciplinary others in the evaluation of the problem and solution to plastic ocean pollution. Nor does it involve critiquing natural science disciplines for ignoring non-experts in local knowledge practices. Loitering forces us to take seriously sociologies contribution as an idiot science to stage encounters with non-experts in urban public settings and open up a space to think. These encounters involve the slowing down of our own disciplinary practices and other's viewpoints. By enabling the opening up of research encounters and the slowing down of thought, the ontology of loitering provoked a re-evaluation of the methods and devices of interviewing for studying engagements with the refill water fountain. In creating an encounter with non-academic others, the ontology of loitering calls attention to the unexpected consequences of social scientists' methods for disciplinary and interdisciplinary problematisations of environmental knowledge practices.

In the following chapter, we pick up on the theme of loitering and idiotic encounters as interactions that intervene in disciplinary and interdisciplinary problematisations. Whereas the current chapter has probed the consequences of accidentally encountered idiots, Chapter 4 demonstrates the advantages that arise from proactively seeking out the idiot as a mode of interdisciplinary research, in what Michael (2022) describes as proactive idiocy. By deliberately introducing idiocy into the research field, we look at how the staging of idiotic encounters with the #OneLess refill water fountain acts as a provocation for activating research methods of observation, attentiveness, and ethnographic description as uniquely positioned to help us engage with how others are provoked to think in problem spaces. Focusing on my attempt to purposefully fabricate an idiot object and idiotic interactions with the refill fountain pilot, we consider the ways the ontology of loitering enables researchers to learn from everyday encounters with mundane water technologies and the publics who engage with them.

References

Agamben, G. (2009) *'What is an apparatus?' and other essays*. Stanford University Press.

Atkinson, P. and Silverman, D. (1997) Kundera's Immortality: The interview society and the invention of the self. *Qualitative Inquiry* 3(3), 304–325.

Back, L. (2012) Tape recorder. In C. Lury and N. Wakeford [Eds.] *Inventive methods: The happening of the social* (pp. 245–260). Routledge.

Barry, A. and Born, G. (2013) Interdisciplinarity: Reconfigurations of the social and natural sciences. In A. Barry and G. Born's [Eds.] *Interdisciplinarity: Reconfigurations of the social and natural sciences* (pp. 1–56). Routledge.

Benney, M. and Hughes, E. (1956) Of sociology and the interview. *American Journal of Sociology* 62(2), 137–142.

Briggs, C. (2007) Anthropology, interviewing and communicability in contemporary society. *Current Anthropology* 48(4), 551–580.

Deleuze, G. and Guattari, F. (1994) *What is philosophy?* Trans. G. Burchell and H. Tomlinson. Columbia University Press.

Foucault, M. (1980) The confession of the flesh interview. In G. Gordon [Ed.] *Power/Knowledge: Selected interviews and other writings* (pp. 194–228). Vintage.

Grieffenhagen, C., Mair, M. and Sharrock, W. (2015) Methodological troubles as problem and phenomena: Ethnomethodology and the question of 'method' in the social sciences. *British Journal of Sociology* 66(3), 460–485.

Law, J. and Ruppert, E. (2013) The social life of methods: Devices. *Journal of Cultural Economy* 6(3), 229–240.

Lee, R. (2004) Recording technologies and the interview in sociology, 1920–2000. *Sociology* 38(5), 869–889.

Lury, C. (2018) Introduction: Activating the present of interdisciplinary methods. In C. Lury, R. Fensham, A. Heller-Nicholas, S. Lammes, A. Last, M. Michael and E. Prichard [Eds.] *Routledge handbook of interdisciplinary research methods* (pp. 1–26). Routledge.

Lury, C. and Wakeford, N. (2012) Introduction: A perpetual inventory. In C. Lury and N. Wakeford [Eds.] *Inventive methods: The happening of the social* (pp. 1–24). Routledge.

Michael, M. (2004) On making data social: Heterogeneity in sociological practice. *Qualitative Research* 4(1), 5–23.

Michael, M. (2011) "What are we busy doing?": Engaging the idiot. *Science, Technology & Human Values* 37(5), 528–554.

Michael, M. (2012) De-signing the object of sociology: Toward an idiotic methodology. *The Sociological Review* 60(S1), 166–183.

Michael, M. (2013) The idiot. *Porto Alegre* 16(1), 71–82.

Michael, M. (2018) Compromising. In C. Lury, R. Fensham, A. Heller-Nicholas, S. Lammes, A. Last, M. Michael and E. Prichard [Eds.] *Routledge handbook of interdisciplinary research methods* (pp. 279–283). Routledge.

Michael, M. (2022) *The research event: Towards prospective methodologies in sociology*. Routledge.

Nordstrom, S.N. (2015) Not so innocent anymore: Making recording devices matter in qualitative interviews. *Qualitative Inquiry* 21(4), 388–401.

Osborne, T. (2013) Inter that Discipline. In A. Barry and G. Born's [Eds.] *Interdisciplinarity: Reconfigurations of the social and natural sciences* (pp. 82–98). Routledge.

Paulus, T., Jackson, K. and Davidson, J. (2017) Digital tools for qualitative research: Disruptions and entanglements. *Qualitative Inquiry* 23(10), 751–756.

Rapley, T. (2004) Interviews. In C. Seale [Ed.] *Qualitative research practice* (pp. 15–33). Sage.

Savage, M. and Burrows, R. (2007) The coming crisis of empirical sociology. *Sociology* 41(5), 885–899.

Silverman, D. (2017) How was it for you? The interview society and the irresistible rise of the (poorly analyzed) interview. *Qualitative Research* 17(2), 144–158.

Stengers, I. (2005) The cosmopolitan proposal. In B. Latour and P. Weibel [Eds.] *Making things public: Atmospheres of democracy* (pp. 994–1003). MIT Press.

Stern, N. (2006) *Stern review on the economics of climate change*. H.M. Treasury.

Whatmore, S. (2013) Where natural and social science meet?: Reflections on an experiment in geographical practice. In A. Barry and G. Born's [Eds.] *Interdisciplinarity: Reconfigurations of the social and natural sciences* (pp. 161–177). Routledge.

Wilkie, A. and Michael, M. (2023) Before the idiot, the poet? Aesthetic figures and design. In M. Tironi, M. Chilet, C. Marin and P. Hermansen [Eds.] *Design for more-than-human futures: Towards a post-anthropocentric worlding* (pp. 103–111). Routledge.

4 Sociography of a curious object

Employing a descriptive methodological agenda for environmental change research that takes seriously the everyday mundane settings of the water fountain, in this chapter we address the role and capacity of descriptive methods as important entry points for social scientists to collaborate with the human and nonhuman participants in the #OneLess campaign to reduce ocean plastic pollution. Based on fieldwork observations of three refill water fountains in London, that took place in three different urban settings for a period of two weeks in October 2018, one week in April 2019, and two weeks in July 2019, we address everyday interactions with the refill fountain pilot. By proactively staging idiotic encounters with the refill fountain, the chapter demonstrates the effects of observation, attentiveness, and description to facilitate engagements with problematisations in the world. Such a sociographical perspective will reveal the role of descriptive research in recognising the everyday problem spaces and embodied practices of refilling as sites for studying problematisation. Investigating what causes thinking in encounters with #OneLess' refill water fountain pilot, the chapter develops new lines of inquiry that broadens the focus of attention from what experts say to direct observations of how others are forced to think in the experimental setting. In paying attention to ordinary modes of problematisation in mundane urban settings, we begin to grasp how knowledge of the environmental problem of ocean plastic pollution is achieved in and through everyday encounters with the refill fountain

DOI: 10.4324/9781003616757-5

Researching #OneLess' campaign pilot project

A question raised throughout this book has been the role social scientists play within global environmental change research. To pick up this question again, in this chapter we consider: What might we learn from placing ourselves in unfamiliar worlds? What might be developed by including the human and nonhuman objects of study in anthropological practices of participant observation? What might be gained by utilising thick description to respond to uncertainty and life as it might become in mundane settings? Almost three decades after the climate crisis, the growing demand for novel academic forms and practices of interdisciplinarity has acquired a new urgency. As we have seen in the previous chapters, the potential for developing new and alternative ways of engaging in environmental research on global change has been a topic of discussion and debate in the social sciences, science and technology studies (STS), and discipline of human geography (Castree 2014, 2015, 2017; Bulkeley 2019). According to the critical human geographer Noel Castree, what's missing from global change research are 'human questions' and approaches to understandings of 'human dimensions' (2017, pp. 57, 63). Outlining the relevance of the people's disciplines for environmental research, Castree elucidates the distance between geoscientists and social scientists as a major obstacle to producing alternative approaches to knowledge practices. In seeking to correct and reverse the paucity of intellectual engagements between geoscience disciplines and the social sciences and humanities, Castree identifies the challenges for broadening the impact of the 'people disciplines' within and outside academia. One of the problems, Castree argues, is that geoscientists ignore and consider irrelevant the work of anthropologists, critical social scientists, and the environmental humanities. A solution, he suggests, is to produce 'more and better interactions' between scientists and specialists in the people disciplines (p. 58). Castree also adds disarmingly, social scientists and humanities scholars have to date 'done a poor job of reaching out to geoscientists or to people outside universities' and 'often remain content to respond to geoscience at a distance within their established disciplinary domain' (pp. 64, 62). The way for scientists, social scientists, and humanities scholars to overcome this distance, Castree argues, is to come out of our professional disciplinary comfort zones and co produce knowledge with a range of stakeholders. Calling for more multidisciplinary research, Castree privileges new means of combining professional expertise 'not merely to collaborate but to *unsettle each other*' (p. 65, emphasis in original).

The late Bruno Latour (2018), in his book *Down to earth: Politics in the new climatic regime*, provides some clues on how social scientist might engage with scientists and a range of non-academic public and professional stakeholders in the co-production of knowledge. In figuring out how to deal with ecological questions sociologically, Latour warns against describing the situation of climate emergency as concerning the publics' lack of understanding of matters of fact and concern. The danger of framing the problem in terms of public knowledge and inventing methods which involve changing people's attitudes, Latour argues, is that we fail to address the 'deficit in shared practice and living in the ruins 'without a shared world' (p. 25). What's at stake, argues Latour, is not a matter of 'learning how to repair cognitive deficiencies, but rather of how to live in the same world, share the same culture, face up to the same stakes, perceive a landscape that can be explored in concert' (p. 25). What this requires in practice, he argues, is not a change in attitude 'but the form and weight of the world to which those attitudes have the function of reacting' (p. 52). In an interview with Jakob Stein and Nikolaj Schultz, Latour points out scientists are 'not withdrawn from the public sphere, but in there, with their special skills, proving with their instruments the existence of the situation' (Stein, Latour and Schultz 2019, p. 219). Without the scientists, the public existence of the situation would not exist, and citizens would not be able to 'absorb the situation' (p. 219.). The sea change 'is when people begin to take it as a question of self-interest' (p. 220). For Latour, what's critical for engaging, motivating, and making a difference is the capacity to connect relations between a people and a land. Other motives he argues make no difference. 'People will simply say "Yes, yes, the planet is in danger, it is not my problem"' (p. 221). In re-assembling the situation in the public sphere and showing and proving to the people the existence of the problem, Latour points out scientists have been instrumental in laying the ground for translating their methods from the lab to land, and their instruments into public practice. The task for social scientists, he suggests, is to study these transformations and describe the reassembly of connections between politics, a land, and a people. The challenge is to '*generate alternate descriptions*' of attachments to local shared dwelling places 'from the bottom up, by investigation', which includes 'the work of description on the part of all animate beings' (2018, pp. 94, 95, 98, emphasis in original). In order to generate alternate descriptions of the composition of local shared dwelling places in the new climatic regime, Latour calls for a methodological shift from critical epistemological analyses of global environmental issues, problems, and policies to descriptions of attachment in local experimental settings. Describing all the beings that participate in the composition of a shared dwelling place is crucial, he argues, as each 'has its own way of identifying what is local and what is global, and of defining its entanglement with others' (2018, p. 93). Drawing on the work of Noortge Marres (2012), Latour stresses environmental change is always an effect of 'object-oriented politics' and is always 'oriented towards objects, stakes, situations, material entities, bodies, landscapes and place' (p. 93).

In this chapter I investigate the methods of the #OneLess campaign team of marine scientists and conservationists to devise and implement the refill water fountain as a local solution to ocean plastic pollution, and the effects of London's refill pilot in specific situations. By attending to how plastic bottled water is articulated as a problem in public spaces and becomes a target of thought and action within the setting of refill water fountain, we learn about the way in which this happens in practice and the necessity of description to bring into focus the objects, material entities, situation, and place that participate in the experiment to prove to people the existence of a problem. The #OneLess pilot project was led by the Marine and Freshwater Team in the Conservation and Policy Department of the Zoological Society of London (ZSL), in conjunction with the Mayor of London to reduce plastic bottled water consumption. The pilot involved the installation of refill water fountains in sites across central London, with 20 installed by the end of 2018. The #OneLess experiment serves as a clear example of how scientists have left the ivory tower and are at the forefront of assembling the public existence of global environmental problems in local public settings. The #OneLess pilot scheme also serves as an example of the relevance of the people's disciplines for environmental research and the role social scientists might play in thinking about how knowledge circulates materially in the Anthropocene. The challenge of broadening the impact of the social sciences to generate descriptions of the #OneLess refill campaign, and connect relations between politics, a land, and people, involved co-producing knowledge with the marine scientists and systems change practitioners at ZSL who led the #OneLess project and stakeholders responsible for maintaining the installation and use of the new refill water fountains. As we saw in the previous chapter, the co-production of knowledge with the #OneLess project team did not involve a straightforward combination of professional expertise, but an unsettling of disciplinary knowledge practices.

The description of the #OneLess campaign that follows draws on my qualitative interviews with the marine scientists and conservationists at ZSL who led the refill water fountain pilot and direct observations of three refill fountains over a period of five weeks during October 2018, April 2019, and July 2019. In my investigation of the refill pilot, I conducted face-to-face interviews with four members of the #OneLess campaign team. My initial encounters with the #OneLess project team at ZSL in October 2018 opened up a series of questions about the methods of the #OneLess campaign and refill water fountain pilot. The question of *how* to get people to value the ocean and build a more ocean friendly society, was central to the public refill water fountain experiment. The goal of reducing the volume of single-use plastic water bottles entering the ocean involved trying to solve the 'missing link' of 'how to connect people with the ocean, and make the connection with people so that they understand how these everyday actions are impacting the ocean'. In addressing the problem of how best to communicate the ocean story and get people to think and make direct connections between the local and the global, the pilot scheme combined a communication campaign to change personal attitudes about single-use disposable plastic water bottles, with an inventive method of creating shareable public refill water fountains. The idea of the pilot was to test if people would use refill water fountains and understand what works and what doesn't to catalyse a behaviour change from plastic water back to the tap. The role of the refill pilot was to enable the #OneLess project team to see how things have changed.

Within the experimental setting of the refill pilot, the mundane technology of the water fountain was deployed alongside visual images of the #OneLess campaign logo inside a plastic disposable water bottle and an image of the ocean to engage the public in environmental participation. The visibility of the hashtag #OneLess bottle in the ocean aims to visually connect people to the global causes and effects of marine plastic pollution and produce public knowledge about marine conservation and ocean conservation. By connecting #OneLess to the problem of ocean plastic pollution, the hashtag marks out a public issue that people could talk about on social media, see what others are saying, and say things themselves. The #OneLess hashtag and mundane technology of the refill water fountain provide a local proactive 'solution' to building an ocean story that is accessible, engaging, and motivating: 'empowering people to do it', and thus produce individual and collective difference around behaviour change. When aggregated together, personal and collective behaviour change in the drinking practices of Londoners and international visitors to the city is seen to make profound impacts on the climate. This local experimental intervention became a 'focal point' for measuring the success of the #OneLess campaign to connect people to a long-term common goal through their everyday choices.

The development of London's refill revolution involved interdisciplinary collaborations between marine scientists, designers, creators, engineers, and architects 'to come up with new ways of doing things'. Bringing designers to work with scientists was seen to increase behaviour change in the experimental setting of the water fountain: 'As a scientist obviously you care about the devil in the detail and so you have to pick the level of engagement and what's going to actually make a difference'. Careful not to approach environmental problems from a large-scale perspective that paralyses people, the #OneLess campaign purposefully avoided demonising plastic or framing plastic pollution in the ocean in terms of economics. As we have seen in Chapter 2, engaging the political capacity of objects, methods, and settings of material participation in environmental issues requires investigating 'how objects become *invested* with specific normative powers through the deployment of particular settings and devices' (Marres 2013 p. 419, emphasis in original). Describing the experimental setting of the eco-show home as an empirical site that asks the confusing question: 'Is it the bathtub or the issue of climate change that draws us in or both?', Marres suggests the example of the eco-show home suspends a singular answer to the question of who or what is doing the engagement (p. 439). In providing a 'public stage' for the 'conferral of capacities of engagement to things' (p. 439), Marres points out environmental experiments in sustainable living offer observations to demonstrate the empirical importance of mundane practices, as well as the active and visible role of material objects and potential of local settings to enable people to 'act upon environmental issues and "be part of the change"' (p. 425). By foregrounding the deployment of instruments to engage residents, stakeholders, and wider audiences including information, posters on walls, doors, and windows and the labelling of objects as experimental methods of material participation, Marres' object-oriented methodology widens the focus of empirical description in the #OneLess campaign to observations of the mundane technology of the water fountain and everyday material objects and visual devices which encourage the public to actively participate in performing environmental change within particular local settings. The descriptive task of specifying the features of material participation empirically is distributed among a broad range of actors and registers. As Marres explains, 'the setting, actors, stuff, statements on posters and the researching theorist all have parts to play' (p. 428). In calling for an expanded empiricism, one which shifts the focus of inquiring from theorising objects and publics to describing the experimental nature of participatory materials 'as dependent on an entanglement with object, devices, and settings', Noortje Marres and Javier Lezaun (2011, p. 503) and Lezaun, Marres, and Manuel Tironi (2016) warn against assuming that the mundane object of the water fountain and devices and settings of the refill experiment simply have normative capacities to solve problems.

Whilst the #OneLess refill water fountain experiment aims to involve the public in environmental issues through mundane practices of material participation, the focus of the #OneLess pioneer network is primarily directed towards engaging a range of private London stakeholders including businesses, organisations, visitor attractions, and venues to help make their transition to one less bottle and make a change. As an evidenced-based organisation, the #OneLess stakeholder evaluation process involves tracking how many companies they work with, how many single-use plastic water bottles these companies have sold, how many they now sell, and how many disposable water bottles are collected from the River Thames along different sites, measured fortnightly. Quantitative data gathered from collecting and counting disposable water bottles is published on the #OneLess pioneer website (onelessbottle.org). Evidence for evaluating the effectiveness of the #OneLess campaign to reduce the number of plastic water bottles in circulation in London was also produced through water flow meters, which recorded how often refill fountains are used and the volume of water going out of them. Deploying statistical methods alongside the method of following the life of plastic disposable water bottles produced evidenced-based evaluations of stakeholder practices. While tracing the life of single-use plastic bottled water from distribution to disposal shows the uses of social research methods for the evaluation of #OneLess' environmental experiment, in my interviews with the #OneLess project team I noticed what was absent from the evaluation process was qualitative data on the mundane technology of the refill fountain and the publics interaction with it. This lack of empirical engagement with public stakeholders and the nonhuman participants in the refill pilot experiment was discussed during my interviews with the marine scientists and conservationists at ZSL and acknowledged by the #OneLess campaign team as an effect of 'grappling with measuring things that are less explicit, more qualitative'. Through conducting interviews with the marine scientists responsible for the pilot, my particular empirical interest in the refill water fountain and experimental setting was made explicit. These initial interviews formed the basis of my collaboration with the scientists at ZSL and informed my subsequent fieldwork in the experimental settings of the pilot. My interactions with the #OneLess project team provoked questions on how social research methods might measure and evaluate the refill pilot, produce data that could lead to an increase in the uptake and usage of the refill fountain, and facilitate the publics interaction with the refill water fountain.

In her description of being and working alongside life scientists, the sociologist Joanna Latimer (2019) illustrates how objects were central for becoming 'entangled' and connected with her scientific colleagues over time and her participation in experimental knowledge making practices (p. 269). Objects were critical for producing research 'openings' and 'possibilities for gathering' (p. 281). What made the creation of common ground between life scientists and a social scientist possible, for Latimer, was understanding '*how* people become attached to and detached from different materials of extension' (p. 269, my emphasis). Paying attention to how and when attachments in common take place, according to Latimer, is not to collapse differences in disciplinary inquiry, or overcome divisions in methodological approaches in the social and life sciences, but to see how our attachments and detachments to and from things brings us into contact with each other and 'reaffirms that which animates and breathes life into our knowledge-making' (p. 269). It is also to do STS 'by other means' that make affective, intimate, and embodied processes of knowledge making with others, including things relevant as an '*object* of but also as a *means* of enquiry' (Latimer and Lopez Gomez 2019, p. 251, p. 252, emphasis in original). In foregrounding intimacy as a site for the social production of knowledge, Latimer (2019) and Latimer and Daniel Lopez Gomez (2019) propose an affective methodology for STS that poses questions about how we become attached. Describing attachments to things we care for and about in interdisciplinary research, for Latimer, 'makes explicit the tensions that would normally keep us apart' (2019, p. 280).

While Latimer's description of her ethnographic experience of being in common with the life sciences involved a form of co-experimenting with intimate human–animal entanglements, my encounters with the marine scientists at ZSL produced different descriptive practices. Coming into contact with the marine scientists and conservationsits who led the #OneLess refill project did not involve an intimate attachment to animals and things mutually cared for and about. The creation of common ground was made possible by not becoming '*intimately* entangled' (Latimer 2019, p. 275, emphasis in original) with *their* attachment to the animal and the ocean and compassion for the welfare of the dead whale, turtle, or sea bird with 'plastic in their guts'. Nor were we gathered by a mutual detachment to single-use plastic water bottles. What made the research opening possible was our *uncommon* attachments. Through my face-to-face interviews with the #OneLess project team at ZSL, the object that made the research encounter opening possible, and brought us into contact with each other's idiotic knowledge making practices, was the marginalised refill water fountain. The process of engaging in a collaborative study of the refill pilot involved grappling with how to measure things that are less explicit and more qualitative in #OneLess' refill water fountain campaign to reduce the amount of plastic bottles in London's waterways and solve the problem of ocean plastic pollution.

Sociography of a curious object

By shifting the object of inquiry from the disposable plastic water bottle to the neglected water fountain and marginalised local experimental setting of the refill pilot, my encounters with the marine scientists brought into focus the particular role of social scientists in a real-world environmental experiment. In making explicit the methodological problem of how to describe and measure the success or failure of the refill water fountain pilot to change Londoner's everyday drinking practices, my participation in #OneLess' experiment opened possibilities for researching environmental issues sociographically. Using social scientific methods involved not only talking to, and following the scientists' knowledge making practices, but also following Anna Tsing, engaging in critical description of the nonhumans ignored and sidelined within the experimental setting of the refill fountain. Taking care to notice undervalued and marginalised things, and how we respond to and engage with them, for Maria Puig de la Bellacasa (2017) requires we remain curious and attentive to *not* knowing about them. What made a sociography of a curious object possible was not allowing myself to judge, or be moved by the refill water fountain, but forced to think by encounters with it. By being there, watching, observing, and noticing the local experimental setting and paying attention to what causes thinking in the publics' encounter with the refill fountain, my goal was to see what could be learnt with others in the empirical setting. In order to topicalise the undervalued and marginalised water fountain, and how problems of plastic pollution are framed by others in questioning situations, I followed Howard Becker's (2016) approach to doing sociology. This involved going into a scene and *clearly* being out of place. Not knowing what to do, or how things work in the water fountain setting, becomes an opportunity for learning. Staying with the discomfort of not knowing how to make sense of the refill pilot, others could see I was out of place and showed me the ropes. In staging an idiotic encounter as a provocation for learning, others attended to the problem I was having and tried to help me through it.

In order to describe other's methods and evaluate the effects of the #OneLess campaign to transform public knowledge of environmental issues, I approached the empirical devices and practical methods of refill in the water fountain setting from the standpoint of a witness. Reclaiming the ordinary ethnographic methods of close observation and thick description enabled me to connect to the ways in which the environmental issue of ocean plastic pollution is experimentalised and apprehended in urban public spaces. Taking hesitation as a starting point to describe what we might learn from the encounter with the refill fountain as a shared event, my fieldwork involved extended periods of looking at the objects, devices, and methods of experimentation from the point of view of their eventual achievement. Participant observation took place in 3 different empirical settings for a period of 2 weeks in October 2018, 1 week in April 2019, and 2 weeks in July 2019. The first fieldwork site involved direct observation at a major central London railway terminus (Figure 4.1).

At first glance, describing this object as utilitarian seems clear cut. The shiny, stainless-steel design is functional and practical. It works

Figure 4.1 #OneLess Pilot Setting.

manually with users having the option of pressing a round silver push-button to deposit water directly into a bottle or leaning in to drink directly from the water fountain. Instructions on where to push and how to use the refill device are inscribed into the steel shell, with images of an index finger pressing a push-button and where to position the bottle. There are no instructions on how to manually use the water fountain's drinking device. While the water fountain's capacity to act and deliver free drinking water to the greatest number of people in a public setting can be measured as a public good, evaluating the activity of this particular mundane object to reduce single-use plastic bottled water consumption is less clear cut. On the one hand, the sign 'Rehydration Point', printed in large bold letters across the top of the refill water fountain and on the floor below printed on a bright blue background of flowing water imagery situates the object as a free water provider. On the other, the setting is saturated with empirical devices of material participation. This includes the signs 'supported by The Mayor of London', the '#OneLess' logo, and a communication digital device that counts the number of 500 ml bottles saved by each use of the refill function, encouraging the public to actively participate in performing environmental change in the local experimental setting. The refill fountain is embedded in a sign that points to it as a 'Rehydration Point', which is big, colourful, and resists ignoring, and becomes elevated to be a refill fountain through the suggested uses it points to, which is done through the signage. In evaluating the success of this empirical setting as an experimental intervention to reduce single-use plastic bottled water, my observations of the refill water fountain took a different direction to Marres' (2013) description of material participation and Marianne de Laet and Annemarie Mol's (2000) descriptive practices with the Zimbabwean bush pump. What caused action was not merely the capacity of the water fountain to alter its boundaries from a solid, clear, and fixed, immutable drinking technology to a fluid, changeable refill fountain, or the empirical devices of material participation, including information about the #OneLess campaign, but how it became this way and worked in this particular practical setting. By unblocking attention to empirical materials *in use*, descriptive research brought into focus the mundane settings and minority practices of refilling.

Description according to Michael Lynch (2013) is critical for engaging what counts as knowledge in the practical settings studied (p. 456). Description can also provide an orientation to how matters of concern 'are composed and invoked by participants' (p. 458). In order to re-orient the practice of description away from theoretical reflections of what's in the world towards ethnomethodological studies of locally situated ontographies, Lynch recommends pursuing ethnographic investigations that describe the mundane, down to earth, uninteresting, and unstudied methods through which members achieve accountable activities and the 'procedures through which that fact is accomplished' (2000, p. 43), issues we explore in greater detail in the following chapter. Ontography, for Lynch, appeals to a notion of the ways in which people construe the shared world (and the basis of it being shared). In paying attention to the ways ontography is a practical lived matter, Lynch's ethnomethodological conception of ontography re-orientates the investigation of the refill water fountain from theoretical reflections of material realities to descriptions of what counts as knowledge in the local setting. Lynch's call for thick description encourages us to examine how participants connect to the mundane technology of the refill fountain in ordinary practices, rather than treating ontology as a matter that can be settled through a priori theoretical discussion.

This move resonates with Isabelle Stengers' (2005) description of an ecology of practices discussed in Chapters 2 and 3. The question of what is knowledge for Stengers is always a practical local problem and 'never a universal problem mattering for everybody' (2010, p. 28). It is a mistake, Stengers (2010) argues, to describe the salience of practices '"as they are"', rather than 'as they become in different surroundings' (p. 26). The challenge for social scientists, is to avoid attributing or imposing a meaning, value, or interpretation to a particular practical setting from 'outside this setting' (p. 26). In order to avoid 'theoretical voyeurism', Stengers proposes social scientists describe the salience of local practices not as they are, or normative and rule following, but 'from the point of view of their eventual particular achievement and what we can learn from it' (pp. 31, 23). What this means in practice, she argues, is learning how to *situate ourselves* in experimentations, not from the standpoint of technology or the majority terms of the 'vested interests of stakeholders' (p. 21), but as a *witness* 'forced to think by the situation' (2018, p. 153). In order to re-direct the focus of inquiry in the experimental setting to the minority practices of laypeople, Stengers calls on anthropologists and sociologists 'not to interpret away' or 'deprive others of their authority about their own practical experience' (p. 150) and accept we are part of those experimentations by playing an active part in apprehending the collective achievements of experimental settings as a minority technique. What this calls for is a shift from descriptions of the ontological multiplicity or singularity of mundane objects (see Woolgar and Lezaun 2013) to a methodological commitment to describe practices in situated encounters as unknown and uncertain from the point of view of their eventual achievement, what Stengers defines as '"ontological tact"' (p. 153). Describing minority practices in experimental settings not as they are but what they *may* become means taking a risk and 'paying attention to the effects of this encounter on you' and 'learning what this encounter demands and how it transforms' (Stengers 2018, p. 154).

The descriptive risk at play in becoming attentive to an ecology of practice that has the power to make us think and feel requires we take seriously nonhuman attachments that force practitioners to wonder and hesitate. The approach ethnomethodologists might bring to such an experimental scene is borne out of everyday members' methods of learning how a setting works by watching what others do

(Lynch 2013), issues we pursue in Chapter 5. Taking hesitation as a starting point to differentiate refilling practices that have the power to force thinking from habitual practices responds to the methodological problem of describing attachments to the mundane technology of the refill fountain and the everyday practical methods of refilling through which members achieve accountable activities in common. By slowing down thought, Stengers method of hesitation asks us to cultivate curiosity to attachments that gather and cause thinking in experimental settings. Creating space for Stengers' practice of hesitation, according to Martin Savransky (2014), is a crucial task for social scientists addressing the challenges of global social science in the age of the Anthropocene. Beginning from the experience of hesitation, Savransky argues, transforms the empirical task to 'make present the need to think about how knowledge circulates materially through the worlds fabric' (p. 244). In order to become attentive to minority practices of hesitation and how knowledge circulates materially, Stengers deploys Tsing's methodological practice of the 'art of noticing' (Savransky and Stengers 2018, p. 144). Noticing, for Tsing (2019), involves ways of looking at human practices, and direct observation of nonhumans and materials, and the way they interact with us from the standpoint of a witness. By being there, noticing and witnessing everyday interactions, and ways of being in a more than human anthropology, Tsing reclaims traditional ethnographic methods of direct observation and thick description as relevant to the ontological turn in the social sciences and STS to 'bring *things* to the center of social inquiry' (p. 240, emphasis in original). In order to expand descriptive practices to the objects of inquiry, Tsing calls for the research methods of close observation and thick description to be taken seriously as tools for noticing the yet to come.

Compared to other empirical sites I investigated, what differentiates the particular experimental setting of the first refill water fountain pilot I observed (see Figure 4.1) is the utilitarian design of the refill water fountain and the visibility of the signage and inscription devices to attach it to an experimental pilot of environmental participation. From an ontographical perspective, this mundane object can be described as both singular and multiple. Public engagement with it is normative and variable. Some people use it to refill their reusable water bottles, others to fill empty single-use plastic water bottles, and those with no empty water bottles to fill, drunk directly from the water fountain. No one used both its refill and drinking functions simultaneously. If people had a bottle, they filled the bottle, and sometimes drunk from the bottle, seeing the primary purpose of the fountain as a bottle-refill station. It also had another practical use as a disposal device to leave waste, including disposable coffee cups, plastic take away coffee lids, and wooden sugar stirrers. The capacity of the refill water fountain as a fluid technology to change shape and shape worlds was less obvious. There was a lot of confusion on what it was, how to engage with it, and ways to make it work. Surprised by the presence of the refill water fountain, I observed people walking past observing it, stopping, noticing, staring, pausing, looking up, glancing at the object as they walked by, watching others using it, stopping to observe and touch, and trying to find out for themselves how it works through trial and error. The success of the empirical setting to transform a disposable drinking culture to an environmentally engaged public involved more than the flexible, adaptable technology of the refill fountain and empirical devices of material participation. What made this refill water fountain work was the uncertainty, doubt, and confusion on what it was, what it was for, and how it worked through shared practices of hesitation and learning. What was salient were the shared practical methods of problematisation composed by the public in common.

Participating in the mundane methods deployed by the public in the experimental setting took a different descriptive turn in my observations of the second refill water fountain I observed at a major historic central London food market (Figure 4.2).

Here there were no visible signs defining this object as a rehydration point. Nor were there any empirical posters to connect the public to the ocean or attach the device to the #OneLess campaign. The refill water fountain's function as an experimental object of environmental participation is ambivalent. What's significant about this particular object is its visibility and oddity as a strange and non-expected object to encounter. What does it do? What is it for? Who is it for? Standing at almost 180 cm tall and 100 cm wide the freestanding outdoor refill water station is elevated by a 10 cm concrete platform. Situated in the centre of the structure is a refill device adjoined to two drinking water fountains, attached at different heights on either side. Combining utilitarian stainless-steel basins and manual pushbuttons to access water with a distinctive dark olive-green colour, the unusual texture and design of this refill and water fountain situate it as a curious

Figure 4.2 Refill and Water Fountain.

object. Its unusual size and shape arouse interest. It's placed very publicly, which displays that this device is to be used by a public. The stage is important also. Individuals can and are intended to be seen to be using this device, which invites others to do so, as they're literally on-stage when doing it. In this way, it is a noticeable act to approach and use this thing. In order to engage with it, participants have to step up onto the elevated concrete platform. The act of stepping up connects the local setting to an experimental context. Being observed and witnessing others in the empirical setting connects the refill and water fountain to a local shared space and common practices of environmental participation. In evaluating the success of this particular device to produce a questioning situation, I became attuned to practices of hesitation in everyday interactions with the refill fountain as relevant for understanding how knowledge of the problem of plastic pollution is achieved. Orienting the practice of description towards locally unstudied settings and ordinary methods produced interesting ontographical observations. While many participants performed refill, and others learnt by watching, observing, and witnessing others refill plastic disposable and reusable water bottles, some drunk directly from the water fountain. In this experimental setting, the ontological multiplicity of the refill and water fountain lost its empirical significance as a problematising object of material participation and became an everyday device to drink from, wash hands and faces, wet hair, wash fruit, clean shellfish, and spot wash clothes.

Hesitation as a practice of knowledge making

Limiting the flexibility of the refill water fountain's working order was central to the third experimental setting I observed in central London. The first refill water fountain installed in the #OneLess pilot project in March 2018, this empirical setting was the most experimental I witnessed. The composition of this particular setting included the familiar material devices of environmental participation located on one tall narrow column (Figure 4.3). The '#OneLess' campaign logo appears alongside an image of a tap and water drop, with the

Figure 4.3 Refill Water Fountain.

inscription 'Refill not Landfill'. Situated in the middle of the column is a small dark green refill fountain, 50 cm high and 10 cm wide, attached to the wall at eye level, with the silver pushbutton below. In the refill area a bright purple bottle-shaped sticker with '#OneLess' logo is visible alongside 'Because everything we do touches the ocean' and the logo 'Sponsored by the Mayor of London'. What was different about this experimental setting was the absence of the utilitarian function of the technology to provide free drinking water to the public. The water fountain works as a refill device only. The design constricts the hydraulic mechanism to a refill function. Attempting to drink water directly from the refill fountain is physically impossible. Trying to access water for multiple purposes is thus practically constrained. Distinctive in design, it's hard to misconstrue its use. While the refill fountain does not look ambiguous, it's not obvious how this curious object works. People have to think about how to get water out of the fountain, why they can't access drinking water, what's different about the way the technology works, and why.

How might we evaluate the success of the #OneLess refill fountain to enable and catalyse change in this questioning situation? How might we measure the capacity of this experimental object to connect the public to the ocean and shape better worlds? While restricting the capacity of the refill fountain to provide free water was deliberately chosen by the management team to prevent the object's mundane use as a water access device, ashtray, and urinal, describing the refill water fountain success through generative constraint-based design cannot be attributed to the majority practices of scientists, designers, and stakeholders involved in the #OneLess campaign to contest plastic water. Empirical observations of what caused the achievement of ontological singularity in this particular setting were not the vested interests of stakeholders responsible for the implementation and everyday management of the refill pilot but the minority practice of participants engaged in practices of hesitation. Reorientating the methodological practice of description away from the standpoint of experts and towards the standpoint of a witness, allowed me to engage in direct observations of the refill water fountain and the practical construction of knowledge in local shared dwelling places. Situating myself alongside the refill fountain, I became attuned to an ecology of practice. Paying attention to the publics' engagement within the experimental setting involved noticing everyday interactions with the refill water fountain and how knowledge circulates materially through practices of hesitation. In my fieldwork, I noticed people stopping and staring, looking up, pointing, glancing and noticing, looking back, touching and pushing, trying and failing to access drinking water, and conversations on how it worked and how to use it. By being there, I witnessed how others are provoked into thinking in their encounter with the refill fountain. This was particularly the case for those who attempted to access drinking water from the refill fountain but failed. What caused action in this particular experimental setting was the provocation of a curious object to arouse collective interest and force thinking. What made the accomplishment of ontological singularity possible in this public setting possible were practices and gestures of hesitation, confusion, and how to respond to uncertainty as a shared collective learning.

In deploying a sociographical methodological agenda for global environmental change, this chapter has shown the role and capacity of descriptive methods as important entry points for social scientists to engage in environmental experiments that enable us to speak back to scientists and do environmental research together. In order to get to grips with the way global environmental problems are assembled in local public settings, the chapter has made visible the problem space of #OneLess' refill water fountain pilot. Engaging in the study of what causes thinking in the experimental setting of London's water fountain pilot, I have shown how matters of concern about ocean plastic pollution are composed and incited in minority practices, and in so doing developed new lines of inquiry for investigating environmental campaigns to contest plastic bottled water. By being fully present as a witness, my empirical investigation of the #OneLess refill pilot has highlighted the importance of descriptive research for taking practices of hesitation as important starting points to join unconnected and previously neglected associations. A descriptive methodological approach gained relevance as a means and method of inquiry to unblock neglected everyday drinking practices and the marginalised object of the water fountain in #OneLess' refill experiment.

As we have seen in these experimental settings, engaging with how a curious object comes to matter not only entails asking questions about the ontological status of mundane technical objects, devices, and everyday settings of material participation in the #OneLess campaign. It also involves becoming attuned to embodied practices of hesitation as central to the problematisation of water. Deploying descriptive methods to evaluate the #OneLess refill water fountain pilot made visible an experiment that turned out to be unsettling and provocative. What this chapter has called for is not a description of environmental change projects but a change in descriptive practices on how others are provoked to think. In the final chapter, we keep the focus on sociographic methods of inquiry as enabling us to stay very close to the water fountain as a zone of problematisation. With an eye to ethnomethodological investigation, we ask what the implications of user's everyday descriptions of the device of the refill water fountain and practice of refilling are, as an opportunity to grasp how others think and point to some of the unresolved issues on how the #OneLess campaign problematises everyday drinking practices.

References

Becker, H. (2016) Making sociology relevant to society. *Journal of Criminal Justice and Popular Culture* 18(1), 1–8.

Bulkeley, H. (2019) Navigating climate's human geographies: Exploring the whereabouts of climate politics. *Dialogues in Human Geography* 9(1), 3–17.

Castree, N. (2014) Changing the intellectual climate. *Nature Climate Change* 4(9), 763–768.

Castree, N. (2015) Geography and global change science: Relationships necessary, absent and possible. *Geographical Research* 53(1), 1–15.

Castree, N. (2017) Global change research and the "People Disciplines": Toward a new dispensation. *The South Atlantic Quarterly* 116(1), 55–67.

de Laet, M. and Mol, A. (2000) The Zimbabwean bush pump: Mechanics of a fluid technology. *Social Studies of Science* 30(2), 225–263.

Latimer, J. (2019) Science under siege? Being alongside the life sciences, giving science life. *The Sociological Review* 67(2), 264–286.

Latimer, J. and Lopez Gomez, D. (2019) Intimate entanglements: Affects, more-than-human intimacies and the politics of relations in science and technology. *The Sociological Review* 67(2), 247–263.

Latour, B. (2018) *Down to earth*: Politics in the new climatic regime. Polity.

Lezaun, J., Marres, N. and Tironi, M. (2016) Experiments in participation. In C. Miller, E. Smith-Doer, U. Felt and R. Fouche [Eds.] *Handbook of science* and technology studies, Volume 4 (pp. 195–222). MIT Press.

Lynch, M. (2000) Against reflexivity as an academic virtue and source of privileged knowledge. *Theory, Culture & Society* 17(3), 26–54.

Lynch, M. (2013) Ontography: Investigating the production of things, deflating ontology. *Social Studies of Science* 43(3), 444–462.

Marres, N. (2012) *Material participation: Technology, the environment and everyday publics*. Palgrave.

Marres, N. (2013) Why political ontology must be experimentalized: On eco-show homes as devices of participation. *Social Studies of Science* 43(3), 417–443.

Marres, N. and Lezaun, J. (2011) Materials and devices of the public: An introduction. *Economy and Society* 40(4), 489–509.

Puig de la Bellacasa, M. (2017) *Matters of care*: Speculative ethics in more than human worlds. University of Minnesota Press.

Savransky, M. (2014) In praise of hesitation: Global knowledge as a cosmopolitical adventure. In W. Kelim, E. Celik, C. Ersche and V. Wohrer [Eds.] *Global knowledge production in the social sciences* (pp. 237–250). Routledge.

Savransky, M. and Stengers, I. (2018) Relearning the art of paying attention: A conversation. *SubStance* 47(1), 130–145.

Stein Pedersen, J.V., Latour, B. and Schultz, N. (2019) A conversation with Bruno Latour and Nikolaj Schultz: Reassembling the geo-social. *Theory, Culture & Society* 36(7–8), 215–230.

Stengers, I. (2005) Introductory notes on an ecology of practices. *Cultural Studies Review* 11(1), 184–196.

Stengers, I. (2010) Including nonhumans in political theory: Opening pandora's box? In B. Braun and S. Whatmore [Eds.] *Political matter*: Technoscience, democracy, and public life (pp. 3–33). University of Minnesota Press.

Stengers, I. (2018) Postlude. *SubStance* 47(1), 146–155.

Tsing, A. (2019) When the things we study respond to each other: Tools for unpacking the material. In P. Harvey, C. Krohn-Hansen and K.G. Nustad [Eds.] *Anthropos and the material* (pp. 221–243). Duke University Press.

Woolgar, S. and Lezaun, J. (2013) The wrong bin bag: A turn to ontology in science and technology studies? *Social Studies of Science* 43(3), 321–340.

5 Using ethnomethodology to make sense of the #OneLess refill water fountain pilot

Continuing to engage the problem space of the London's refill water fountain pilot, this chapter addresses how plastic bottled water becomes articulated as a problem in the practices of experts responsible for the #OneLess campaign and the publics who interact with the water fountain. Using the interview technique, the chapter compares the system change perspective of #OneLess professional practitioners involved in the design and evaluation of the refill pilot with the voices of people engaged in the use of the refill fountain as part of their everyday practices. Drawing on qualitative data gathered from 4 interviews I conducted with marine scientists and conservationists, and #OneLess practitioners at the London Zoo (2018–2019) and 130 interviews with refill water fountain users in 3 central London sites over a six-month period (2018–2019), the chapter compares the methods of systems change with the practical methods people use to make sense of the refill fountain in common. The aim of examining expert and common practices of knowledge making alongside each other is to investigate what we learn about the #OneLess project from users everyday methods in the situated locations of the refill experiment. In giving a voice to and thus taking seriously how problematisations are articulated in thought and practice by research participants engaged in the use of refill water fountain, we take seriously the activity of refilling as an accountable and thus recountable inter/action (Garfinkel 1967/1984; Sacks 1992).

Focusing on the inter/action between the refill water fountain and the mundane work of refilling as a methodic apparatus to render social phenomena explorable and analysable, we see what studies in ethnomethodology contribute to understanding the impact of the

DOI: 10.4324/9781003616757-6

#OneLess water fountain experiment and the role of Foucault's concept of problematisation as a resource to articulate how thinking with water happens. That is, by making explicit the interaction between the refill fountain and common sense making practices of users as a practical concern, the chapter renders the ordinary methods of refilling explorable and analysable. In so doing, we consider what an ethnomethodological study that takes as its starting point the mundane activity of refilling contributes to the evaluation of #OneLess' refill pilot. In teasing out the distinctive contribution of the sub-discipline of ethnomethodology to engage the practical, situated ways people make sense of the refill fountain, the chapter provides a framework to understand how Isabelle Stengers' (2023) reclaiming of common sense as a shared sense of a questioning situation operates in practice through shared ways of thinking and understandings. More specifically, studying the activity of refilling as an interactionally produced action and speech act in situ, the chapter elaborates on what ethnomethodology offers for investigating environmental experiments in sustainable living as ordinary practical problematisations. By making visible how plastic water is articulated as a problem 'in and as of' the everyday settings of the refill water fountain pilot (Garfinkel 2002, p. 92), we learn how ethnomethodological practice and writing shape data in ways that allow the making, interpretation, and translation of findings as an interdisciplinary problematisation.

The practitioners perspective – leveraging change

From the #OneLess practitioners perspective, what's required to eliminate single-use plastic water bottles entering the ocean involves creating systems change at the level of policy, infrastructure, and culture. Anna Birney (2021), Director of the School of Systems Change responsible for the #OneLess campaign, uses the concept of the lever to describe the steps deployed to help build and grasp ideas of how change might happen. Using the mechanical metaphor of a lever, Birney outlines the various steps required to bring about a systems change to create a refill culture across the city of London. This includes where to intervene, how to get to grips with the practical resources necessary to work with change, understanding what impact would be, and how to measure it. The key leverage points identified by Birney for bringing about systemic awareness of the problem of plastic ocean pollution involve infrastructure design to support refilling, communication campaigns on valuing the ocean to become an ocean friendly society, creating behaviour change by tackling negative perceptions about tap water, and changing social norms to make single-use plastic water a taboo. In seeking to change things by reconfiguring the system from within, the #OneLess campaign sought to rebuild the infrastructure of water fountains in London through new design solutions and social norms that drive publics to drink water from a tap rather than a plastic bottle.

The #OneLess systems change perspective invites several questions which we address in this final chapter. The systems change way of thinking implies the public resist behaviour change due to a negative perception about the drinkability of tap water. A further assumption is that people's participation in the refill pilot is caused by the narrative of how the ocean is framed in communication and policy. The claim that changing social behaviour is expressed in terms of a sympathy with the ocean and sea life implies an ethical superiority over those who engage in the continued use of plastic bottled water. The presumption that these deeper leverage points would impact behaviour in practice treats ocean messaging and valuing the ocean as the drivers to effect change and the refill fountain as a minor entity in the process of environmental participation. In so doing, the systems change perspective marginalises the role of the refill water fountain in the #OneLess project to hydrate Londoners without single-use plastic. While acknowledging the transformative impact of deeper leverage points of mindset and everyday practice are difficult to devise, implement, measure, and articulate, Birney argues 'the less obvious, but potentially more effective and powerful areas of intervention' (2021, p. 763) require finding indictors that see impact and change. In practice, this demands 'working with change, rather than trying to box, quantify or make static the way we view impact' (p. 763).

By seeking to create change in the system of hydrating Londoners through *leverage points*, the practitioners' perspective does not underestimate the role of articulating how systems change can be achieved through minor interventions. What informs the articulation of catalysing change in the #OneLess refill experiment is transforming how the narrative of the ocean is framed in communication campaigns. The central leverage point through which to frame the problem of plastic pollution and engage people in the solution to drink from the tap involves shifting the narrative of the ocean from its economic value to social value, as well as recognising the ocean's contribution to well-being. The methods deployed by #OneLess practitioners to leverage small shifts in public awareness of the ocean include communication campaigns involving social media, visuals, and the #OneLess hashtag. Other infrastructural solutions used to promote ocean optimism involved working with designers to address physical obstacles and barriers to using public water fountains, such as hygiene, and feeling comfortable to drink tap water and access refill stations in urban settings. #OneLess' design fellowship worked with product designers, service designers, and graphic designers to apply design and innovation to address problems of how to create change in London's drinking practices. The aims of the design fellowship were to 'scale infrastructure digitally and physically', 're-imagine and renew infrastructure', 'enable spontaneous access to water', and 'make refilling the new norm'. Motivating people to make an 'ocean connection' and 'link everybody in their choice to the ocean' is articulated as an outcome of the success of messaging communication and design campaigns to 'directly connect people with cause and effect' and change thinking and values. These leverage points are understood as a central mechanism to produce behaviour change.

In my interviews with the marine scientists, marine conservationists, and #OneLess practitioners, I was probed on how a sociologist could contribute knowledge of behaviour change. My interdisciplinary collaboration with the #OneLess project team was assumed to involve 'looking at social norms and how they've changed over time, right?', helping turn them into points for leveraging interventions. However, as part of thinking beyond the self-imposed limits and spaces of possibility discussed above, through my interviews I was able to take into account the claims of the systems change perspective to provide solutions to the problem of plastic ocean pollution and develop another way to engage in the process of problematisation in #OneLess' experiment. My goal was to affirm the material design and communication devices central to the #OneLess refill pilot without sidelining and ignoring the interactivity between everyday practices of refilling and the mundane setting of water fountain.

What emerged from initial interviews about the #OneLess project was the relevance of social research to examine the claims regarding the leverage points of systems change. From the system change perspective, ocean messaging campaigns are seen as one of the most effective interventions in creating change in public perceptions and drinking behaviour. As a place to intervene and solve the issue, messaging campaigns situate the 'ocean at the heart of all communication'. The tag line 'everything we do touches the ocean', which appears on all communications and posters involves 'connecting the one less bottle initiative back to the wider global plastics agenda'. The #OneLess communication approach 'gets people to value the ocean' from a 'more positive angle', rather than from a 'fear perspective'. The messaging campaign 'one less bottle in the ocean' was devised as an alternative to the way the environmental problems of the climate and plastics crises are represented as a 'huge problem where people are paralysed by the scale of it, and can't actually do anything that's going to have an impact, because it's so big'. The hashtag ocean optimism is used to balance the narrative of 'doom and gloom' and provide solutions. The #OneLess communication campaign is perceived as central to addressing the 'question of how can we get people to work together and value the ocean' and build a more 'ocean friendly society'. The use of visuals, including the #OneLess hashtag, are seen as leverage points, creating a 'direct connection with cause and effect'. By providing accessible solutions and making them easy to use, the aim of #OneLess communication campaign is seen to empower people by giving them hope in a context of solastalgia; a feeling of being 'hopeless about the future of the environment and just thinking about how the past used to be'. #OneLess' messaging logo 'gives each person a sense of personal agency to make change themselves' and 'ensure one less bottle is going in the ocean', thus making it a 'common issue to solve' and giving the ocean narrative 'a global purpose'. What matters is:

> trying to enable change by connecting people to their deeply held values and putting the ocean at the centre of the campaign, so that people then feel connected to the ocean, and because of that, then they are more willing to protect it.

The use of representational figures of wildlife endangered by plastic marine pollution in James Honeyborne, Mark Brownlow, and David Attenborough's documentary *Blue Planet 2* (2017) is understood as having 'put the issue of marine plastic pollution on the radar'. Visual representations of entangled turtles and sea birds with plastic in their guts became 'an issue that got traction and engagement'.

From the #OneLess practitioners perspective, the refill water fountain becomes a focal point to measure the effects of ocean messaging campaigns to change thinking and values and produce 'a common goal to solve'. The distinctiveness of leverage points as a lens to frame problems and solutions to ocean plastic pollution lies in their capacity to change mindsets and social values. The refill water fountain is situated as a key indicator to evaluate the *effects* of media and messaging campaigns to communicate the issue around the ocean and marine protection. The role of the refill pilot is to evidence the *impact* of design solutions and messaging campaigns to target people's values around the ocean and the issue of ocean plastic. The use of the refill water fountain is measured as an *outcome* of transformed thinking that creates new habits of behaviour to drive a refill drinking culture. While the refill water fountain is seen to act as both a 'trigger point' for a particular action to be taken and an object to measure social change, the impact of thick and thin leverage points in making changes to Londoner's everyday drinking practices gives little attention to the affordances of the mundane technology and setting of the refill fountain and how change happens in practice. The language of triggering reveals the constrained field of vision that the system change paradigm imposes on itself. That is, the systems change practice of proactively mapping and scaling up solutions too quickly tries to look past its own starting points: the process and practice of thinking with water and the problematising device of the refill water fountain as a potentially shared ground in enacting a mindset and behaviour change. The impact of ordinary situations and everyday practices to problematise thinking are left unaddressed. Rather than suggesting that the category of leverage points should be superseded, in my interviews with the #OneLess project team I suggested it might be more productive to examine the refill fountain as an empirical device and find ways of understanding how change in Londoners' drinking practices happens in experimental settings by inventing ways to describe and measure it. This involved investigating #OneLess' communication and design strategy through the lay meanings attached to the ocean messaging campaign and the design solution of the refill fountain in everyday practice.

Using the 'environ-mental imagination' as an entry point for examining 'environmental relations to identity, place, practices, and nonhumans', Kathryn Yusoff and Jennifer Gabrys (2011, p. 521) call attention to the role of humanities scholars and practitioners in collaboration with scientists. Recognising the changed conditions of knowledge production of climate change as an imaginative act requires a methodological approach that joins the sciences, social sciences, and humanities in interdisciplinary collaborations. Addressing *how* communication practices operate, Yusoff (2010) and Yusoff and Gabrys (2011) draw into focus the creative method of environmental imagining as a critical leverage point to consider the ways knowledge of the problem of marine plastic pollution is engaged with and experienced in the #OneLess campaign. In Yusoff's work, iconic images of endangered animals are perceived as a 'prosthetic emotional device' that 'extend our sensory capacity to both notice and respond to environmental change' (2010, p. 76). The practice of seeing and witnessing representations of entangled turtles and sea birds with plastic in their guts force viewers not to look away, but look back in ways that disrupt our thinking and redistribute our senses to reform our politics and practices (Yusoff 2010). In this sense, Yusoff and Gabrys (2011) show why it may be useful to explore how iconic objects and images in the popular imagination mediate the climate and plastics crises for audiences.

Focusing on the object of the teapot in environmental campaigns, advertising, news, and publicity media, and how everyday objects are represented to help protect the environment, Noortge Marres (2014) also provides a way to conceptualise objects, as well as methods to analyse how ordinary things become charged with a range of problems and acquire political capacities as a result. Marres' empirical investigation of the politics of the environmental teapot is informed by John Dewey's (1955/1908) concept of problematisation. Taking her cue from Dewey's analysis of *problematic situations* that require resolution, Marres points out 'there is nothing resolved, or neat or fixed about a politics of objects' (p. 266). Quoting Dewey, Marres suggests that in order to account for the politics of objects, we must pay attention to the '*problematisation of things*':

> 'valuation takes place only when there is something the matter; when there is some trouble to be done away with, some need, lack or privation to be made good, some conflict of tendencies to be resolved by means of changing existing conditions'.
>
> (Dewey 1955 (1908) cited in Marres 2014, p. 266, emphasis in original)

For Marres, studying the problematisation and issuefication of things as something that plays itself out through objects involves textual methods of issue analysis in different media settings, including online textual and visual analysis.

Evaluating Marres' contribution to developing empirical forms of inquiry for understanding environmental problems, Andrew Barry (2021) draws attention to what he sees as the limitations of a Deweyan notion of the problematisation of things. Contrasting Marres' Deweyan conception of problematisation (as a negative valuation of a trouble, lack or matter to be made good) to Foucault's positive account of problems, Barry points out:

> Problematisation doesn't mean the representation of a pre-existent object, nor the creation through discourse of an object that doesn't exist. It's the set of discursive or nondiscursive practices that make something enter into the play of the true and the false, and *constitute it as an object of thought.*
>
> (Foucault 1989, p. 296, cited in Barry 2021, p. 95, my emphasis)

The task of giving empirical substance to environmental problems, according to Barry, is to show how research may generate multiple responses from experts and lay publics about solutions that may not be singular and produce novel effects. What this means in practice, Barry argues, is that we should not expect environmental experts and other interested parties to have a 'common interest in a particular outcome' (p. 108). Nor should we treat environmental problems as singular and universal. What concerns Barry, is that science and technology studies' (STS) theoretical and empirical account of problematisation and issuefication are 'too restricted in their empiricism' (p. 105). Marres' Deweyan-inspired object-oriented commitment to an articulation of the process of problematisation, according to Barry, limits the analysis of problems to 'the study of readily observable traces and texts' (p. 99). By limiting the focus of inquiry to published documents and texts, including those that can be traced on the internet as sources of data, Barry argues Marres' empiricist approach to controversy analysis raises questions as to 'which problems are not articulated as objects of concern, as well as which voices may be excluded in the process of problematisation' (p. 99).

In seeking to expand the empirical programme of research on environmental problems, Barry calls for social researchers 'to be more empirical, by recognising the need to interrogate the limits of what is readily observable and discernible'. This includes 'what can and cannot be problematised, what can and cannot be rendered public, and what voices and arguments have and have not been heard' (p. 112). What an empirical programme of research requires is forging new kinds of interdisciplinary research and collaborations between natural and social scientists, as well as between experts and non-experts. Proposing a sociological approach to the study of how problems are generated and transformed in practice, Barry directs the researcher to 'be sceptical of the claims of those who know in advance what the solution to the problem is and of those who know what hidden forces have determined what has happened' (p. 113). From Barry's sociological perspective, the marine scientists and systems change practitioners involved in the #OneLess campaign are considered too close to the articulation of the problems and solutions to ocean plastic pollution, and STS theorists engaged in a Deweyan conceptualisation of the problematisation of things too distant from mundane processes and ordinary practices of problematisation to fully analyse it. A social scientific standpoint informed by Foucault's mode of problematisation is seen as better placed to navigate environmental questions and problems with both experts and non-experts. The empirical challenge is finding ways to engage with particular situations and 'demonstrate the value of rigorous abductive inferences that do not necessarily lead to determinate solutions but, to the contrary, might come to pose further problems' (p. 113). Barry's sociological insights prompt us to look again at the significance of problematisation in the #OneLess refill experiment, noting that studies of the environmental problems have been locked into a theoretical approach to problematisation, at the expense of more thorough empiricist approaches. But in advocating an abductive methodology, for both expanding links between Foucault's proposition on problematisation and the task of showing 'how research may generate multiple solutions and, indeed, further problems' (p. 95) that differs from what is expected based on theoretical understandings, Barry does not expand on the tools required for such an empirical investigation.

By bringing the work of STS into conversation with Harold Garfinkel's (1967/1984) studies in ethnomethodology, Marres (2012) goes some way to addressing the empirical challenge of investigating problematisations within particular urban experimental settings. In her analysis of how experiments in sustainable living can be explored empirically, Marres likens sustainable living experiments in the contemporary context with the research technique of Garfinkel's breaching experiments in the 1960s. Garfinkel's breaching experiments involved social scientists deliberately modifying ordinary routines and activities of social life. These disruptions were produced in order to provoke actors to generate accounts of seen but unnoticed everyday objects, activity, and action. Highlighting the similarities between these breaching methods and sustainable living experiments in the contemporary context, Marres points out breaching is no longer an intervention enacted by ethnomethodologists in the world but a disruption of everyday ways of doing things by a range of social actors and devices. The purpose of social research, Marres argues, is to engage in the task of 'adding its techniques and modes of research and analysis to those already deployed', so that social researchers 'may render or foreground different aspects' of what is being problematised and 'produce accounts diverging from those in circulation' (p. 90). Through such participative ways of working empirically, the sociologist is seen as having a potentially critical role to play in 'describ[ing] the multiple ontologies articulated in practices of "living environmentally"' (p. 90). Responding to Marres' question of how social researchers might participate productively in breaching experiments of sustainable living initiated by others outside the university, in what follows we consider what studies in ethnomethodology and the methodological repertoire of ethnomethods contribute to an investigation of problematisation in #OneLess' refill water fountain pilot.

By making trouble in ordinary situations, Garfinkel's (1967/1984) research aim was to disrupt everyday micro-level interactions and make the ordinary strange. The objective of producing dis-organised interaction as a starting point to understand society from within was to observe the ordinary methods people put to use to make sense in common of the social world. Recasting social researchers as participants in the practical methods people use to make knowledge in common, Garfinkel's (1967/1984) methodology de-privileges method as a device that social scientists deploy to understand the social. Being surprised by ordinary everyday methods of common understanding of the social world, Garfinkel's (1967/1984) intervention challenges social researchers to reflect on our taken-for-granted assumptions about the social and test sociologies thinking imagination. Studies in ethnomethodology for Garfinkel, Michael Mair (2025) explains, are not a reparative enterprise but investigations in the role language plays in our practices and as our practices. Explaining what ethnomethodology does, Mair points out that for Garfinkel instruction comes from engagement with particular structures of practical action and reasoning. The key to ethnomethodology is found not in theoretical texts but 'in the practices in and through which we together produce worlds known and knowable in common' (p. 92). Given the methodological focus of the discipline, it is not surprising that the most sustained application of Garfinkel's ideas to the social world has occurred in the social sciences and STS. Along with Mair (2025), Lucy Suchman (see Suchman, Gerst and Kramer 2019) and Michael Lynch (2011) have shown a particular commitment to demonstrate not only that methods should be treated as topics for ethnomethodology, but the influence of Garfinkel's distinctive style of investigation in which trouble was a central methodological theme. As regards to Garfinkel's original breaching experiments, Lynch (2011) refers to the use of practical difficulty as methodologically influential for STS, as 'leverage for revealing taken-for granted practices that perform what a "fact" or "thing" is' (p. 32).

The challenge posed by Garfinkel's studies in ethnomethodology for social scientists and researchers engaged in understandings of global environmental issues is not reducing the study of social action to external causes. For Suchman (see Suchman, Gerst and Kramer 2019), understanding Big Issues requires paying close attention to the body, practices of embodiment, and material practice in the mundane reproduction of everyday activity. Studying the enactment of social order in everyday mundane activities involves not taking the place, things, interactions, and embodied relations of sense making for granted. It also involves having to 'find the sites in which those categories are being produced and reproduced, and then actually see how that's been done, materially' (Suchman, Gerst and Kramer 2019, p. 10). This means it becomes harder to identify knowledge of the problem of ocean plastic pollution outside of the material settings of #OneLess' refill pilot and the shared or at least potentially shared ground alluded to earlier. Ethnomethodology is particularly useful for generating common sense understandings of refilling practices because it obliges social scientists to question the conventional understanding of action and meaning by #OneLess systems change practitioners and STS researchers as separate processes and insists on the importance of meaning making as a practical matter and not a theoretical one 'internal to and resolved in social life' (Mair and Sharrock 2020, p. 76). Significantly, ethnomethodology's distinctive concern with the methods people use in doing social life involves 'understanding concrete instances of practical action and reasoning, ordinary life's work, in and across a multitude of settings from within' (Mair 2025, p. 92). As this quote suggests, despite its insistence on taking ordinary methods seriously, ethnomethodological studies require recognising the practical grounds of inquiry, which make it difficult to appraise its capacities for dealing with #OneLess' refill water fountain campaign without paying close attention to meaningful activities of refilling in ordinary mundane practices and everyday local settings.

By taking Garfinkel's interest in analysing mundane interactions as key for understandings of the #OneLess refill fountain pilot beyond a thin description of leverage points to an ethnomethodological study of refilling in situ, the next section of this chapter switches focus from the theoretical and methodological perspectives of the #OneLess systems change approach to the public actors engaged with the refill water fountain. The aim of engaging in an ethnomethodological investigation of the refill fountain is to open up possibilities for descriptions of the world in research practice and address the potential for social scientists to productively contribute to existing environmental experiments. In producing ethnomethodological work in practice, I offer a descriptive approach that seeks to evaluate the leverage points of #OneLess' systems change perspective in situ. Evidencing an ethnomethodological study of refilling in action, as an intervention for linking together and engaging otherwise disparate and ignored material objects, devices, and practices of making sense of the refill water fountain pilot, we might find ourselves in a better position to consider what everyday observations and descriptions offer for questioning received views and getting to grips with how plastic bottled water is made a question and problem of thought in ordinary drinking practices.

The participants' perspective – refilling as an ordinary method

By taking the site and setting of the #OneLess refill pilot and the publics' situated activity of refilling as a topic of inquiry, is it possible to think again about what social scientists might add to this environmental experiment? Studying users' descriptions of their methods of refilling, the remainder of the chapter presents an empirical case study of refilling in specific public settings in London. Using examples drawn from fieldwork in two urban locations, I discuss the everyday methods of engagement with the objects, settings, and devices of the refill water fountain. Examining the interpretation of ordinary activities of refilling by participants in mundane practices through Garfinkel's ethnomethodological programme involved documenting 'the embodied practices through which persons singly and together, retrospectively and prospectively, produce *account-able* states of affairs' (Lynch 2000, p. 33, emphasis in original). To investigate the practical action of refill in ordinary conversation, I used in situ co-located interview research at the refill water fountain as a topic to engage non-specialist explanations and study the phenomenon of refill in 'the local procedures through which members "achieve" account-able activities' (Lynch 2000, p. 43).

Doing ethnomethodological studies involved the use of locally sited interview questions that were developed in dialogue with the marine scientists and systems change practitioners at Zoological Society of London (ZSL). This involved developing shared vocabularies of data collection. In order to produce qualitative research on the refill water fountain pilot that would be seen as useful and relevant data through an interpretation of findings that could be shared, we co-produced a series of questions. The research questions sought to address how the meaning of refill is made through rational and practical reasoning: *Is this the first time you've used the water fountain? What made you use it today? Have you used others? Have you seen the #OneLess campaign? What's your motivation for using the refill station? Can you think of anything that would improve it or any obstacles that make it difficult to use?* By focusing on the ordinary methods used by members of particular pilot settings, we gain a fuller appreciation of common sense making activities. The interview questions address how actors in different settings identify problems in those settings, understand and use the refill water fountain, interpret their own practice, think about data on the #OneLess campaign, and are motivated to action.

In asking questions about user's practical involvement with the water fountain and knowledge of the #OneLess project, we don't take the meaning or practice of refill for granted. Making strange the ordinary activity of refill draws out troubles faced when practically engaging with the water fountain. The questions about practical action and reasoning draw attention to the more mundane aspects of drinking water in daily life overlooked by #OneLess practitioners. These interview questions test assumptions about the systems change perspective, particularly taken-for-granted assumptions about the leverage points of design and communication messaging strategies, including ocean optimism and the #OneLess hashtag as triggering action. Incorporating questions on practical action corrects the limits of established ways of conceptualising data in the methods and practices of systems change practitioners and carries us more deeply into the minutiae of knowledge formation in different empirical settings of the refill water fountain pilot.

By interviewing users in different practical settings, I was able to get as close to the material methods, actions, and interactions of sense making as possible. Studying knowledge formation in three urban settings alongside the refill fountain, bodies of refiller's, and others enabled an investigation of the local production of sense making in which the social is participatively rendered observable and recordable in situ. Paying more and closer attention to embodied practices of refilling was not about evaluating whether refilling is done, or done correctly, but seeing the activity as already in the hands of experts, who are for Garfinkel 'masters of their own methodologies' (Lynch 2011, p. 935). Addressing in more detail the practical actor's description demonstrates the relevance of lay explanations of mundane activities for understanding problematisation as an observable intelligible phenomena. The interview transcripts, collected from 130 refill water fountain users across 3 sites over a 6-week period, highlight how knowledge of an environmental problem of plastic pollution is practically achieved through embodied everyday practice. Using examples drawn from 2 sites which feature the #OneLess campaign logo, including the hashtag and ocean messaging of one less bottle (as shown in Figuers 4.1 and 4.3 in Chapter 4), we see how knowledge of refill is an accomplished practice of mutual participation and mutual intelligibility, produced through co-presence with others (Garfinkel, Rawls and Lemert 2006). Drawing on interview transcripts provides a practical understanding of problematisation in action, grounded not in terms of theoretical concepts, political objects, norms of valuing the ocean or an environ-mental imagination, but co-participation in a shared method of producing a questioning situation, that is seen, witnessed, and observed by others (Garfinkel, Rawls and Lemert 2006).

The following examples are chosen to show how refill is undertaken through the methods not of systems change but of the practical everyday activity of water fountain users' methods, that are reportable and observable. These exemplars, draw attention to meaning making practices in situ. The three examples are selected because they stand as examples of views held in common amongst my research participants in their descriptions of the refill fountains. The chosen exemplars draw attention to general findings across the different field sites. In terms of broader insights, the vast majority of interviewees did not mention taste or water quality as an obstacle to using public water fountains and overall had very little if any knowledge of the #OneLess campaign. Whilst many participants did talk about not wanting to use more plastic as their motivation for using the refill water fountain, the environmental harms of ocean plastics pollution to marine life was not something they discussed. By adapting ethnomethodological research techniques to everyday refill practice we gain an insight into the significance of the unstudied concrete methods and situations around which people gather and what is achievable at the smaller scale.

The examples show how refilling is accomplished and the practical work that underpins it. The three transcriptions draw attention to aspects of refilling that are passed over in the system change account of #OneLess practitioners. The transcripts produce evidence of mundane practice in which the everyday doings of refill water fountain users are not left out. The users' description in ordinary language of practice, evidence a different more common sense understanding of refill. The following descriptions consist of a shared activity of practical inquiry with others engaged in an encounter with the refill fountain in a local public setting. While 'live data' can sometimes be treated as pre-eminent within ethnomethodology, this way of working leverages the point that we learn of the world, our place, and relations within it by talking to people and following them in what they do alongside us. This use of a shared and shareable ground, to give it another of its multiple senses, is and has to be locally methodologically adequate, as the work of Garfinkel and Harvey Sacks amply demonstrates (see Mair and Sharrock 2020).

Texts 1, 2, and 3 demonstrates how refill is done. Transcription is employed as a method to exhibit practical reasoning. It provides an account of a set of practices in which refilling is enmeshed. The transcripts present embodied practices of noticing and witnessing the refill fountain as it happens, as ways of knowing in situ. Conversation with users in situ reveals the way the practice of refilling and object of the water fountain come into being by making sense to others through language of the social setting. Using interviewing to capture and present actors' observations and descriptions of the settings and details of refilling, we see how lay practitioners are involved in revealing problematisation as an ordinary accomplishment.

Text 1

Interviewer: Is this the first time you've used a water refill fountain in London?

Respondent: Yes, it is.

Interviewer: And what made you come and use this one?

Respondent: Partly because it's free so you can get a refill for free, and partly because it means you don't have to buy another bottle. So you're not using more plastic. So you can reuse an existing bottle of plastic.

Interviewer: What motivates you to carry that refill bottle around with you?

Respondent: To have some water. But also, because I don't want to keep buying bottles, and actually don't want more plastic in the environment.

Interviewer: Ok, and how did you find the design of this water fountain?

Respondent: It's good generally, yes it looks fine. The only general thing I had was working out how to actually get the water to come out the tap. To start it wasn't immediately obvious. But I mean yes after a few seconds, we worked out how.

Interviewer: Have you been aware of any campaigns going on across London about refill culture?

Respondent: I haven't, no I haven't seen anything of that nature.

Through the process of encountering the refill fountain and being interviewed about it, practical action in the setting and refill culture is made strange. Experts show how practical reasoning of the refill water pilot is worked out – where the process of problematisation becomes witnessable as ways of seeing and thinking that are accountable to others. Observation and interaction become ways of making sense of one less in common. Context-dependent sense making activities formulate ideas of what making sense of one less bottle in the ocean is and what is involved in making sense of the refill water fountain pilot. Making sense of refill culture with others takes place not through representations of endangered animals, ocean friendly communication campaigns, or the #OneLess hash tag but through embodied mundane practices of engagement: 'To start it wasn't immediately obvious. But I mean yes after a few seconds we worked out how', where the intelligibility of both the initial problems and their resolution with the water fountain is imminently worldly.

Text 2 demonstrates how sense making is done in actions and words. Refill operates in the use of language to craft a world and understanding of the water fountain. This is observable when I ask experts to describe details of their use of the refill fountain. Words are descriptive of practices of seeing and doing. Following J. L Austin (1962), words are also a form of doing. Words demonstrate competence in knowing how a fountain connects to the environmental issue of plastic pollution. Knowing is invoked by the existence of the signage plus the interview question. The words do the work of describing, but they also do other work. The interview technique became a resource for encountering practical action and making everyday activities observable in language. Language as a sense making activity produces a set of instructions of what members are doing and why. Their accounts have the capacity to be read instructionally by others. In treating members' talk of refill as instructive, we also acknowledge that displaying order to newcomers is part of what's being done here. Talking draws my attention to material mundane practices to help me understand user's methods. Listening to the accountability of user's actions becomes a guide to finding action through words and how we recover action and its meaning in context.

Text 2

Interviewer: I noticed that you stopped to use this water fountain today and you have a refill bottle.
Respondent: Yes.
Interviewer: Do you carry it around with you all the time?
Respondent: Correct.
Interviewer: And is this the first time you've used this particular water fountain?
Respondent: Yes, it is, we tried the one up there, but it's broken.
Interviewer: What made you notice this one?
Respondent: It's not broken and it's next to the train station.
Interviewer: Ok, so you just spotted it?
Respondent: Yes.

Interviewer: Have you found other fountains where you can refill?
Respondent: No.
Interviewer: What do you think of the design of this one?
Respondent: It's easy to use and you can fill a whole water bottle, and my daughter has to drink 2 litres of water a day, so perfect.
Interviewer: Any way it can be improved?
Respondent: Probably even more signage so people can see it, we only saw it at the last minute, so that would be helpful.
Interviewer: What's your main motivation for carrying around these refill bottles?
Respondent: Obviously a) environment, b) the amount of fluid we need to drink, and c) I don't actually like going to the shops to collect more plastic package.

Here we learn how problematisation does not pre-exist practice as an abstract phenomenon. By locating action inside the refill setting where encounters with the water fountain and others takes place, the respondent guides us to see the ways in which problematisation as a practical action is articulated and made accountable as an embodied everyday practice of knowledge making. Language describes how technology is put to work in everyday problem spaces. Practical concerns with how to use the fountain produce action. Refilling as an everyday activity is accomplished through social relations in the encounter within the water fountain setting. Mundane practices of seeing, observing, and pressing buttons bring a new thing and phenomena into being.

Text 3 is a productive account of how lay experts link things together in encounters with the refill fountain. The transcript shows how interactions with the water fountain enable refill to happen, and through this joint work, give it whatever stability it comes to have. Through sharing methods, members reveal how they build a social world through sense making activities. Refill operates through bodily practices of witnessing, observing, noticing, finding, approaching, touching, and working out. The scenic relevance of sight and practical visibility of the water fountain are central to practices of doing refill. Corporeal co-presence and the possibilities it opens up is more important than communication campaigns. Here we see how embodied engaged activity gains more credence in the user's description of everyday practices of refill than the visibility of the #OneLess hashtag. Describing proximity to the refill water fountain, 'noticing' it, it's visibility in space, and sociality with it, draw attention to the lay practitioner's method of witnessing as a key practice of doing refill. Knowing how to use the water fountain is interactionally managed in the space of the refill setting through observation and following others, including the fountain's lead (cf. Pollner 1979).

Text 3

Interviewer: Is this the first time you've used this water fountain?
Respondent: It is, yes, it is.
Interviewer: What made you notice it today?
Respondent: Well, I was just walking past and just going to get the train and then I just saw someone else come and use it, and I was really surprised that it was here. Very excited, because I've been wanting one for quite a while.
Interviewer: Have you seen them before?
Respondent: This is the first time I've ever used one in London I think actually.
Interviewer: Ok.
Respondent: Yes, I haven't really noticed them before. I noticed one before in another station actually, but I didn't need to fill up, and this is the first time. I'm back and forth from X quite a lot. It's the first time I've noticed it here, so I don't know how long it's been here for.
Interviewer: Have you seen the #One Less campaign to minimise plastic bottled water?
Respondent: I have yes, the campaign yes.
Interviewer: So you have a sense of refill culture?
Respondent: Yes, people have become more aware of it, and like they bring their coffee cups just to use it. People are becoming more conscious about it.
Interviewer: What's your motivation for using this refill station?
Respondent: Well definitely to not use as much plastic, because when I normally come to London I just come for the day. I feel like I'm constantly buying water bottles. Obviously trying to drink as much as I can, I will constantly buy lots of water bottles. I am conscious that I'm ending up using quite a lot throughout the day

which doesn't feel good. So, I really am glad there's more of these little water fountains here.

Interviewer: Can you think of anything that would improve them, or any obstacles that make it difficult to use them?

Respondent: I can't think of any, just maybe for them to be a bit more obvious that they're there. I don't know how long that one's been there for.

Interviewer: And what would make it more obvious?

Respondent: I don't know, maybe if it was just like in the middle there, so you would see it when you walked past. I literally only just saw it, because someone was walking past, and had just come from using it. So otherwise, I wouldn't have noticed it. Interesting to know how long it's been there for, because I'm always at X, as I say.

Interviewer: Any problems that you associate with water fountains or using them?

Respondent: No, none at all, it's great. I wish there were more, definitely.

Text 4 demonstrates what ethnomethodology offers social scientists and practitioners for evaluating #OneLess' refill experiment. Here we see a sense of focus on embodied sense making practices, a concern with mundane practices in particular settings, and a descriptive approach to practical detail. This is observable in a debriefing session at the London Zoo with a member of the #OneLess project team which followed the completion of my interviews with research participants. In the following interview, I report back my analysis of the data I gathered from interviews and observational research conducted on the refill water fountain pilot. The transcript shows how interdisciplinarity is produced through the interview interaction and shared interpretation of data. It also draws attention to the interview technique as a strategy of problematisation to make everyday ordinary activities observable and accountable to professional others.

Text 4

Respondent: I'm interested to learn what you've found from your analysis of the pilot. If that's alright?

Interviewer: Yes, that's why I'm here. I wanted to tell you all about it. I looked at 3 different sites. What I would say for all 3 refill fountains is that it's about seeing, seeing is key. I asked participants 'have you heard about the [#OneLess] app', [they answered] 'no', 'do you know about the [#OneLess] campaign', 'no'. The people I interviewed in general don't know about those things. So, what's bringing about the doing of refill is the knowledge created there and then in that space. It's a relational shared space that's produced by the object of the water fountain and that's where knowledge is produced, that's where change happens. It's a seeing, doing, thinking experiment. No one is coming to me and saying oh yes, I have all this knowledge about the #OneLess campaign and the problem of plastic in the ocean, and yes, I have access to the app and I am here to produce the solution.

Respondent: So, it's a trigger point?

Interviewer: You could call it a trigger point, but something about the water fountain matters. This object you're looking at, especially people clocking it, they're looking at it, and they're looking at other people, and what they're doing with it. The thinking and the doing are happening in that time and space, and it's important in terms of [location] X where you've got such a big space. People are walking through the space, they're noticing it, some of them said they'd gone into the shop next door and seen the refill fountain and come out and used it. Even if people aren't stopping to use it, I would say that is a space where people are seeing it, where people are walking past, observing, noticing.

Respondent: So, it's almost like they're participating in their own act of enquiries. They're seeing this refill fountain and then they are, it's like their own little, oh this is different. It's like their own little mini action enquiry is happening, triggered by as you say, seeing this refill object that they don't know what it is, this alien object in a lot of cases.

Interviewer: That's the key thing. We don't look at anything unless we've noticed it, and we only notice it because we've observed it, and we've observed it because it's disrupted our everyday routine. So, a big part of the #OneLess experiment is that the refill water fountain is disruptive. As soon as something is disruptive, then you're able to get people to think about their everyday practice and what they are doing.

Respondent: Ok, that's really interesting.

Interviewer: For the majority, it's mostly their first time. They've not seen the #OneLess app, and don't know about the #OneLess campaign. 'What made you use it'? 'I just saw it' and 'I needed water'.

Respondent: Was that based on the interviews that you did or from observations?

Interviewer: Mostly from speaking to people. Almost everyone I interviewed said it pretty much. Very few recognised the #OneLess campaign. Hardly anyone knew about the app, but many of my interview respondents talked about plastic, so they did make that connection. It wasn't just that this is free water, it's accessible, and people are going to save money. That came up a lot. It was important but so was doing refill. A lot of people are refilling single-use plastic bottles.

Respondent: I was going to ask if there were any negatives? It sounds like on the whole it was very positive, and you received some very positive comments.

Interviewer: The #OneLess pilot is successful not because it disciplines people to do things in a particular way. It works because it gets people to think about something that's going on in the world, by seeing this local response to a global problem. Obviously connecting people to the ocean is a wonderful thing, and the local is also part of it. The #OneLess systems change strategy is to get businesses involved. But it's also about people, they're key. People are really important, especially how they think about the refill pilot, how they make sense of it, how they create knowledge with it. That's what I found interesting. The interviews I did here at ZSL [with the marine scientists] were fascinating because they described what the #OneLess project was about, why you were doing it, and how you were going about it. But the evaluation of the campaign from a systems change perspective isn't really public focussed. I think it was X who said we need you [a sociologist] to understand what people think of them [the water fountains]. We need to know whether there are better locations to put them, and we need to create data to generate new ideas that will get people to commit to wanting more of them. There's location, and of course there's architecture, there's engineering, there's design, there's communication campaigns, there's all of these things that practitioners talk about in terms of leverage points to make change. What's different or perhaps what I add to the project, is the people's perspective. What are people doing, how are they thinking about the water fountain. It's not necessarily like [they are saying] I have knowledge, I act on my knowledge, and my knowledge causes and determines my actions. I'm there [studying] refilling practices in the world so we can understand people's response to that experiment. Refill fountains are an important part of the process. How people access them, their behaviour, their mindsets, their perceptions, their values. The question of how do they and how do we think about it, describe it, and determine its success, that's a big question. How do we go about evaluating the campaign in terms of the message about the ocean and plastic pollution. I found that in the moment you disrupt people's routine background expectations, you're creating a problem space where people have to think about what they're doing, and why, and how they're doing it. This is a shared question, that's relational, and social.

Respondent: This is really interesting for me, in terms of the seeing as doing. The seeing is what leads to the doing, and it being a process of action enquiry for the participants themselves. So, I guess my question is, looking forward in terms of what you know, how can we make these objects more accessible to more people. How can we increase usage, how can we have this uptake. Do you think there's any role for putting the fountains on Google Maps, for example? Or do you think there's any role for a wide scale campaign across the city or is that just not going to work? Is it all about that moment in time when you see the water fountain. Thinking in terms of future interventions that might help to make these objects more available to increase uptake, in your mind does this have potentially any impact?

Interviewer: I would say yes install more. People want more, that's what they say, 'there should be more'. People don't say 'oh we should get of rid them', 'why are they here', no-one's said that.

Respondent: But do you think raising people's awareness about the fact that there are more helps? Your kind of research has shown that it's very much people see them and then they use them, they don't necessarily come prepared with their bottle. That's not how it works based on your research. If we were to intervene and inform people that there are 20 fountains, do you think that would help, based on the results of your research?

Interviewer: I think that's something to consider. Having more of them available will create more knowledge. I found people were walking by and seeing and observing and witnessing others and being observed themselves. And the whole of [location] X is participating in that event, either directly or indirectly.

Respondent: Ok, that's really interesting. It's really interesting you mentioned there's quite a few different distinctions in terms of interaction. I was wondering whether you feel you can comment on whether the aesthetic impacts use, and whether there was any trend in terms of if there was more usage at different fountains. I'd be interested to know of design.

Interviewer: I think that's an important question about the different settings. There's something about seeing the refill fountain, and the shared, common, mutual participation in knowledge making that's happening in that particular space, maybe because of the aesthetics, design and visuals, or just because people are there.

Respondent: I guess my question is, are there any interventions we can make. I'm thinking about this as a kind of change maker. How can we increase usage? What could we do to help to facilitate that interaction? What would be helpful? Is there nothing we can do? Is it all about people just walking past?

Interviewer: It's not just about installing fountains. It's also the interactions going in and around those spaces that matter. We have to make sense of one less in common, and it's important that it's situated, and it's interactive, and it's a part of sociality. I don't think we can separate these issues from the design and communication campaigns.

Respondent: It's great that it has that wider impact in terms of making common sense of the plastics agenda. It's interesting for us that very few passers-by said that they knew about the campaign. In terms of awareness, as in triggering them to approach the fountain, they seem to just be doing it because they're there.

Interviewer: Yes, that's what I'm finding, it's occasioned.

Respondent: So, if we leverage more time into raising awareness, is that going to be a trigger? That's my question. These fountains for us were an experiment, and it's really fascinating to see you participating in that experiment, and learning from what you've found, from your involvement in it. But is that your specific area of interest the fountains, or for example, I know you have seen our work on the design challenge. I'm just wondering in terms of whereabouts it is in what we do that interests you and your observation of it all?

Interviewer: I'm interested in what people are doing at the fountain, and that's not something that you are necessarily, but it's part of your experiment. It's important but it's not your primary focus.

The transcript is relevant as it enables us to see how the methods of the systems change practitioners including the design and visuals become understood as different from the mundane methods of lay members. Understanding the specific ways practical methods of refilling matter is interpreted by the #OneLess practitioner as a topic of interest. By making ordinary common sense scenes and common sense making activities a source of data, the research interview enabled me to give a voice to the unnoticed methods and ordinary language of problematisation described by participants. By making the mundane methods and ordinary language of refill visible, the systems change practitioner learns about the locally occasioned and thus locally encountered problem space of the refill fountain as involving shared methods of co-producing a questioning situation that is witnessed, seen, observed, and acknowledged by others. Seeing is not a metaphor for vision, but an activity that involves different ways of 'seeing' – glancing, noticing, observing, scrutinising, scanning – which open up the practice of seeing as a way of doing refill (Coulter and Parsons 1991). In listening to my findings of everyday refill practices in the mundane settings of the pilot, the practitioner acknowledges what's novel methodologically about an ethnomethodological investigation of the #OneLess campaign.

By elaborating on why and how the refill fountain matters as an empirical device to realise the development of problems and solutions to plastic ocean pollution, the transcript also shows the reader what researching, thinking, and writing ethnomethodologically enables for investigating everyday problematisations and provides methodological guidance on how to study it. Here, we heed the council of Mair (2021) who maintains that in order to grasp the challenge of Foucault's work for research 'we must abandon the programmatic commitment to problematisation' and investigate 'what sense is made of the problem in practice' (pp. 104, 149). In this closing section of the book, we have witnessed the relevance of ethnomethodological studies as a practical resource to investigate #OneLess' refill experiment. Showcasing the approach of ethnomethodology for investigating the #OneLess campaign has demystified the process of conducting interdisciplinary research and demonstrated the significance of problematisation for evaluating the methods of systems change in environmental experiments. Examining the system change perspective of #OneLess practitioners responsible for the design, implementation, and evaluation of the refill water fountain pilot, alongside the ordinary methods of research participants engaged in the practice of refilling, we have seen the difference ethnomethodological and ethnographic data collection practices make for engaging the ordinariness of problematisation. Studying mundane methods, as an apparatus to render problematisation explorable and analysable in the settings of the water fountain, further reveals how the interview technique, participant observation, and detailed transcription contribute to the elaboration of the #OneLess pilot project as particular situated problematisation.

Through a study of ethnomethodology in action, we have learnt about the distinctive contribution of ethnomethods for connecting the voices of systems change practitioners, professionals from diverse disciplines, and publics. In showing how to use ethnomethodology to research an environmental experiment in sustainable living, the chapter has also reflected on the strengths and blind spots of the theoretical and methodological approach of systems change. In pulling together how knowledge of the water fountain has been generated through different epistemologies and methodologies, the chapters in this book have demonstrated what the case study of the refill fountain reveals of disciplinary differences and what could be done differently with Foucault's notion of problematisation. Moreover, I hope to have encouraged readers to be inspired by the interdisciplinary approach offered by my empirical case study of the public refill water fountain and inquire into how participating in ordinary methods that make people think might be put into practice.

References

Austin, J.L. (1962) *How to do things with words*. Oxford University Press.

Barry, A. (2021) What is an environmental problem. *Theory, Culture & Society* 38(2), 93–117.

Birney, A. (2021) How do we know where there is potential to intervene and leverage impact in a changing system? The practitioner's perspective. *Sustainability Science* 16(3), 1–17.

Coulter, J. and Parsons, E.D. (1991) The praxiology of perception: Visual orientations and practical action. *Inquiry* 33(3), 251–272.

Dewey, J. (1955 [1908]) Theory of valuation. Reprinted in O. Neurath, R. Carnap and C. Morris [Eds.] *International Encyclopaedia of Unified Science* 2(4). University of Chicago Press.

Foucault, M. (1989) The concern for truth: An interview with Francois Ewald. In S. Lotringer [Ed.] *Foucault live* (pp. 293–308). Semiotext(e).

Garfinkel, H. (1967/1984) *Studies in ethnomethodology*. Polity Press.

Garfinkel, H. (2002) *Ethnomethodology's program: Working out Durkheim's aphorism*. Edited with an Introduction by A.W. Rawls. Rowman and Littlefield.

Garfinkel, H., Rawls, A. and Lemert, C. (2006) *Seeing sociologically: The routine grounds of social action*. Routledge.

Honeyborne, J., Brownlow, M. and Attenborough, D. (2017) *Blue planet II*. BBC Earth.

Lynch, M. (2000) Against reflexivity as an academic virtue and source of privileged knowledge. *Theory, Culture & Society* 17(3), 26–54.

Lynch, M. (2011) Harold Garfinkel (29 October 1917 - 21 April 2011): A remembrance and reminder. *Social Studies of Science* 41(6), 927–942.

Mair, M. (2021) *The problem of the state*. Routledge.

Mair, M. (2025) Theoretical orientations: Editorial: Ethnomethodological readings of philosophy, social theory and the social sciences. In M. Mair [Ed.] *Routledge international handbook of ethnomethodology* (pp. 91–103). Routledge.

Mair, M. and Sharrock, W. (2020) Action, meaning and understanding: Seeing sociologically with Harvey Sacks. In R.J. Smith, R. Fitzgerald and W. Housley [Eds.] *On Sacks: Methodology, materials and inspirations* (pp. 19–31). Routledge.

Marres, N. (2012) Experiment: The experiment in living. In C. Lury and N. Wakeford [Eds.] *Inventive methods: The happening of the social* (pp. 76–95). Routledge.

Marres, N. (2014) The environmental teapot and other loaded household objects: Reconnecting the politics of technology, objects and things. In P. Harvey, E. Casella, G. Evans, H. Knox, C. McLean, E.B. Sliva, N. Thoburn and K. Woodward [Eds.] *Objects and materials. A Routledge companion* (pp. 260–271). Routledge.

Pollner, M. (1979) Explicate transactions: Making and managing meaning in the traffic court. In G. Psathas [Ed.] *Everyday language: Studies in Ethnomethodology* (pp. 229–255). Irvington.

Sacks, H. (1992) *Lectures on conversation*. Blackwell.

Stengers, I. (2023) *Making sense in common: A reading of Whitehead in times of collapse*. Translated by T. Lamarre. University of Minnesota Press.

Suchman, L., Gerst, D. and Kramer, H. (2019) 'If you want to understand the big issues, you need to understand the everyday practices that constitute them'. Lucy Suchman in conversation with Dominik Gerst and Hannes Krämer. *Forum: Qualitative Social Research* 20(2), 1–18.

Yusoff, K. (2010) Biopolitical economies and the political aesthetics of climate change. *Theory, Culture & Society* 27(2-3), 73–99.

Yusoff, K. and Gabrys, J. (2011) Climate change and the imagination. *WIREs Climate Change* 2, 516–534.

Index

abductive problematization, practice of 18
Actor Network Theory (ANT) 17, 29, 34, 48
A future history of water (2019) 7–9
Anthropocene, problem of 56, 91, 101
anthropology 8, 70–1, 101
Attenborough, David 113
Austin, J. L. 125

Bacchi, Carol 41
Back, Les 76
Badiou, Alain 54
Ballestero, Andrea 5, 7–9
Barnett, Clive 18, 22, 36, 40–1, 60–1; understanding of problematisation 42
Barry, Andrew 60, 66–7, 72, 116–17; engagement with Foucault's work on problematisation 12–13, 16–18, 78
Becker, Howard 97
Beck, Ulrick: analysis of climate change 54; theory of metamorphosis 54–6
Birney, Anna 110–11
Blue Planet 2 (2017) 113
Bodies of water: Posthumanist feminist phenomenology (2019) 7
Born, Georgina 66–7, 72; engagement with Foucault's work on problematisation 12–13, 78
bottled water industry 1, 50
bottled water vending machine 32, 39, 43
Bridge, Gary 18, 22, 40–1, 60–1; understanding of problematisation 42
Briggs, Charles 78
Budds, Jessica 5, 30, 32, 39

Castree, Noel 89
Chilvers, Jason 14–16
Chen, Cecilia 6

de Laet, Marianne 99
Deleuze, Gilles 19, 72
devices: concept of 77; Ballestero's definition of 8; Law and Ruppert's definition of 78; Lury and Wakeford's inventory of 77–8; *see also* tape recorder, as a distinctive sociological device
Dewey, John 115; analysis of problematic situations 115
'Do Something' water fountain campaign (2009) 51, 53, 61; Hawkins et al.'s description of 59–60
Down to earth: Politics in the new climatic regime (2018) 90
drinking fountain *see* public drinking fountain

ecology of practice, concept of 57
Ekers, M. 22, 35–7, 40
emancipatory catastrophism 55
environmental humanities 5, 89

ethnomethodology 24, 51–3, 93–4, 99, 117–23, 134–5

Foucault, Michel 6, 31, 56, 116; Barry and Born's engagement with the work of 78; notion of problematisation 12–13, 16–17, 19–20, 24, 29, 36, 39–45, 49, 57, 61, 67, 79, 110, 117, 135; notion of the dispositif and apparatus 77–8

Gabrys, Jennifer 114–15
Garfinkel, Harold 24, 51, 117–23
Gomez, Daniel Lopez 95
Gramsci, Antonio 35–7
Grieffenhagen, Christian 79
Guattari, Felix 19, 72

Hamlin, Christopher 1–2
Haraway, Donna 56
#OneLess campaign (London, 2016– 2022) 2–3, 10, 13, 16, 18, 20, 23, 44–5, 47, 61, 65–7, 74, 83, 94, 96, 99, 102, 109–13, 117, 120–1; communication and design strategy 114
#OneLess design fellowship 112
#OneLess logo 99, 105, 113
#OneLess Pilot Setting 98
#OneLess pioneer network 94
#OneLess professional practitioners 109, 112, 133–4; at the London Zoo 109; perspectives of 110–20
#OneLess refill water fountain project 66, 79, 86, 88, 91, 104–6, 109, 114; evaluation of 110; role of the refill water fountain 111; social research on 89–96
#OneLess stakeholder evaluation process 94
#OneLess water fountain experiment 110–11, 117
Hawkins, Gay 22, 38–40, 47–51, 58–9, 61; evaluation of the technique of problematisation 59
hesitation, as a practice of knowledge making 104–7
history of water 35
hydrolectics 31–2; Linton's method of 35, 37–8, 42
hydroscience 30–3
hydrosocial cycle 30–2, 43; Linton's delineation of 34, 37–8

idiot, in social science 72–5
Interdisciplinarity: Reconfigurations of the social and natural sciences (2014) 12–13, 23, 66–71
interdisciplinary environmental research 65; agnostic-antagonistic mode of 68–9, 71; integrative-synthesis model of 67–8; problematisation of 66, 67–72

Kaplan, Martha 43–4
Kearnes, Matthew 14–16
Kierans, Ciara 60
Kirby, Vicki 6
Koopman, Colin 42

Landstrom, Catharina 5, 11–14
Latimer, Joanna 95
Latour, Bruno 29, 43, 90–1
Lee, Raymond 78
Lezaun, Javier 13–14, 94, 100
Linton, Jamie 5, 30; ANT relational approach to waters 34; case study of the water fountain 31; delineation of the hydrosocial cycle 34; effort to grant nonhumans a role in the historical analysis of water 33–4; Foucauldian-inspired genealogy of water 29, 33; historical analysis of thought in relation to water 43; hydrosocial research paradigm 22, 29, 31, 37, 41; method of practical hydrolectics 33, 35; practice of analytical hydrolectics 31–2, 42; *What is water?* (2010) 29–30, 33–4
Liboiron, Max 2
Loftus, Alex 22, 33–4, 36–7
loitering, as an interdisciplinary method 84–6

Lury, Celia 77–8
Lynch, Michael 99, 101, 119, 121–2

MacLeod, Janine 6
Mah, Alice 2
Mair, Michael 60, 79, 118, 119
Marine CoLABoration 3
Marres, Noortge 13–14, 51–3, 91–4, 115–18
metamorphosis, theory of 16–17, 54–7, 61
Michael, Mike 57–9, 66, 72–5, 81, 85; account of the links between the idiot and the discipline of sociology 81; reflections on the uses of the idiot in social science 72–5; an ecology of futures 57–9
Mol, Annemarie 99
Mythen, Gabe 54–5

naturecultures 6
Neimanis, Astrida 5–7
Nordstrom, Susan 79

object-oriented methodology 94
object-oriented politics 91
ocean plastic pollution 16, 92, 96, 110, 119, 131; environmental problem of 13, 23, 88, 97; impact on marine life 123; #OneLess campaign to reduce 88
ontography 99
ontological tact 100
Osborne, Thomas 18–19, 60, 66, 69–71; assessment of sociologies contribution to interdisciplinary problematisation 71–2; engagement with Foucault's work on problematisation 18–20, 60; methodology of problematology 18

parasitism, academic practice of 72
pharmakon, metaphor of 55–6
plastic bottled water, consumption of 3, 91; contestation of 10; environmental campaigns to reduce use of 13, 53; Hawkins and Hawkins et al.'s case study of 38–9
plastics pollution 2, 97, 131; global environmental issues of 4, 21, 92
Plastic water: The social and material life of bottled water (2015) 12, 22, 48, 51, 58
platonic interdisciplinarity 71
poaching, academic practice of 71, 72
posthuman feminist phenomenology 7
Potter, Emily 39, 47
proactive idiocy 85–6
problematisation of things 115–16
public water fountains 3, 4, 43; in Canada 32; 'Do Something' water fountain campaign (2009) in Sydney 51, 59–61; investigation of in New York 43–4; in London 98, 102–5, 111
Puig de la Bellacasa, Maria 97

'Quvrez un grand cu' education campaign (France, 2011–2015) 10

Rabinow, Paul 42–3, 60
Race, Kane 22, 39, 47
refill water fountain 74, 77, 79–80, 88, 95, 104, 110, 121, 127; installation of 91; role in #OneLess project 111; *see also* #OneLess refill water fountain pilot 98–9, 102
Rose, Nikolas 18–20, 60

Sacks, Harvey 109, 123
Savage, Mike 18, 20, 60, 76
Savransky, Martin 16–18, 57, 101
Schultz, Nikolaj 90
science and technology studies (STS) 5, 8, 48–9, 65, 89, 95, 99, 101, 115–16, 119; Marres' empirical analysis of material participation 51–3, 93–4
sensory politics 9–10
Sharrock, Wes 79, 119, 123
Shove, Elizabeth 48–9, 58

single-use disposable plastic water bottles 3, 50, 92, 94, 96–9, 101, 104, 110, 111
social science 16, 41, 58, 66, 70, 101; research methods 14–16, 58, 71, 75–8, 81, 84–6, 89, 90, 94–7, 101, 106
Spackman, Christy 9–10
Stein, Jakob 90
Stengers, Isabelle 11, 16–17, 22, 55–7, 100, 110; concept of an ecology of practice 57; cosmopolitan proposal 72, 73; definition of Deleuze's figure of the idiot 72; engagement with Foucault's notion of problematisation 16–17, 56–7; *In catastrophic times: Resisting the coming barbarism* (2015) 55–6; practice of compromise in interdisciplinary research 73; practice of hesitation 101
Stern Report (2006), on the economics of climate change 70
Strang, Veronica 10 Suchman, Lucy 119
sustainable living experiments 51–3, 65, 117–18, 135
Swyngedouw, Eric 34, 54–5
systems change perspective 109–13, 117, 119, 120–2, 130–5
tape recorder, as a distinctive sociological device 77; advantages and disadvantages of 76; Back's analysis of 76–8; as an idiotic object 75, 77
The taste of water: Sensory perception and the making of an industrialized beverage (2023) 10
Thinking with water (2013) 6
Tironi, Manuel 94
trespassing, practice of 72
Tsing, Anna 97, 101

Urry, John 55

Wakeford, Nina 7
water fountains *see* public water fountains
What is water? (2010) 5, 29–30, 33–4, 37
Whatmore, Sarah 11–14, 66, 68–9
Whitehead, Alfred 5
Wilkie, Alex 53, 73

Yusoff, Kathryn 114–15

Zoological Society of London (ZSL) 3, 65, 95, 121; Conservation and Policy Department 91; #OneLess project team at 96